Beyond the Grave with AI

Interviews with Famous Figures

BORKO FURHT

AF697634

Copyright © 2025 Borko Furht (Boca Raton, Florida, USA)

All Rights Reserved.

Table of Contents

Part III — Leadership & Faith: Voices That Moved Nations and Souls

Part IV — Courage & Spirit: Champions of Will and Determination

Beyond the Grave with AI: Interviews with Famous Figures

Preface

There are questions that time itself never answers. What would Einstein think of our quantum age? Would Marilyn Monroe have found peace in today's world of endless fame? How would Muhammad Ali face an opponent made not of flesh, but of code?

These are the whispers that inspired this book.

As a lifelong researcher in artificial intelligence, I have spent years teaching machines to analyze, predict, and understand. But I have always wondered — can AI help us *feel*? Can it help us connect with the humanity that shaped our past?

This idea haunted me until it finally turned into a challenge: to bring together history and imagination, reason and wonder, technology and soul.

Before I began, I asked my AI assistant a simple question: *Is it right to use AI to interview the dead?* Its answer was thoughtful — that such an endeavor could be meaningful, as long as it remained transparent and respectful. That was enough for me.

And so I embarked on a journey across centuries, guided by algorithms and human curiosity. I spoke — through the lens of AI — with twenty remarkable figures who shaped our civilization: from scientists who bent the laws of nature to artists who painted emotion itself, from world leaders who faced history's storms to athletes who defined courage.

Each conversation is imagined yet inspired by truth. Every response is shaped by what these individuals said, wrote, and believed while alive. The result is not a resurrection, but a *reflection* — a dialogue between what once was and what still echoes within us.

This book is not just about AI. It is about our eternal longing to understand greatness, to hear wisdom one more time, to bridge the silence between life and memory.

—Borko Furht

Boca Raton, Florida

Introduction

Artificial Intelligence has given humanity a new voice — one that can echo the past and speak to the future. In *Beyond the Grave with AI*, that voice becomes a bridge across time. Through it, I sought to listen to those who once defined human thought, beauty, and courage — not as ghosts, but as living presences in the digital imagination.

Here, Albert Einstein contemplates the mysteries of existence beside Nikola Tesla, whose dreams of light still hum through our world. Marilyn Monroe speaks candidly of fame, loneliness, and love. Pablo Picasso reflects on the rebellion of art. Muhammad Ali's fierce rhythm turns into poetry, and John F. Kennedy looks back on hope and unfinished work. Even Pope Francis, though still among us, joins this dialogue that stretches beyond mortality — a voice of faith in the age of machines.

These are not interviews in the journalistic sense. They are conversations in the realm of *what if* — poetic reconstructions of thought, guided by data, biography, and intuition. They live somewhere between reality and reverie, where imagination honors truth and technology becomes a mirror for memory.

I did not set out to prove that AI can imitate genius. Rather, I wanted to see if AI could help us *reconnect* — with wisdom, with empathy, with the fragile beauty of human imperfection. In these pages, AI becomes not a machine of logic, but a vessel of curiosity, helping us rediscover the souls of those who built our civilization.

Every interview is an invitation: to listen, to wonder, to remember.

Because the past is never truly gone — it waits for us, patiently, beyond the grave.

Albert Einstein: An Imagined Dialogue

Introduction (Narrator)

He was more than a scientist—he was a symbol of genius itself. Albert Einstein, born in 1879 in Ulm, Germany, forever changed the way humanity understands the universe. With his theory of relativity, he redefined time, space, and gravity. With his voice, he spoke for peace, justice, and human dignity in an era scarred by war and extremism. Yet behind the equations and the fame stood a man—curious, playful, stubborn, and deeply human. In this rare interview, Einstein reflects not only on his groundbreaking science, but also on his philosophy of life, his struggles, and his vision for the future of mankind.

Albert Einstein (14 March 1879 – 18 April 1955) was a German-born theoretical physicist who is best known for developing the theory of relativity. Einstein also made important contributions to quantum theory. His mass–energy equivalence formula $E = mc^2$, which arises from special relativity, has been called "the world's most famous equation". He received the 1921 Nobel Prize in Physics for his services to theoretical physics, and especially for his discovery of the law of the photoelectric effect.

Q: Can you explain to "normal people" what your theory of relativity is?

Albert Einstein: My theory of relativity is about how space and time are connected. Most people think of time as a steady tick-tock, the same for everyone, and space as a fixed stage where events happen. But I discovered that space and time are not separate and not fixed — they form a single fabric we call *spacetime.*

If you move very fast, close to the speed of light, time for you slows down compared to someone who is standing still. Distances also change depending on how you move. This is called *special relativity.*

Later, I realized gravity is not a mysterious pulling force, but the bending of this spacetime fabric. Imagine putting a heavy ball on a rubber sheet — the ball makes a dip, and smaller balls roll toward it. That's how the Sun bends space and time, and that's why planets orbit it. This is *general relativity.*

So, in short, motion changes time and space, and gravity is the shape of space and time. The universe is not a stage — it is alive with curves and warps, and we are all moving within it.

Q: It is still too complicated. Could you please reframe your answer as if you were speaking to a child with a very simple analogy?

Albert Einstein: Imagine you're sitting on a train. If the train goes super, super fast — almost as fast as light — then your clock inside the train will tick slower than the clock of someone standing on the ground. To you, everything feels normal, but to the person outside, your time looks like it's moving in slow motion!

Now, imagine space and time are like a giant trampoline. If you put a bowling ball on it, the trampoline bends. If you roll a marble nearby, it doesn't go straight — it curves around the bowling ball. That's what gravity is! Big things like the Sun bend space and time, and that's why Earth goes around it, just like the marble.

Q: Do you believe that physics will ever find a "final theory" of everything, or will there always be new mysteries?

Albert Einstein: I have always believed that nature is governed by a deep order, a harmony that the human mind can uncover. Physics is our attempt to write down this harmony in symbols and equations.

Will there be a 'final theory'? Perhaps. I once sought what I called a *unified field theory*, a single framework that would describe both gravity and the forces of atoms. But the more we learn, the more subtle nature appears.

Maybe one day science will find a complete and elegant theory. But I suspect that every answer will open new questions. The universe is like a great book: even if we read the last page, we will still wonder who wrote it, and why.

But if we ever think we've found the last word, nature will surely whisper another riddle in our ear.

Q: Which of your contributions do you personally consider the most important — relativity, quantum theory, or something else?

Albert Einstein: Relativity may be the work for which I am best known, but I sometimes feel my early papers on the quantum nature of light were equally important. They opened the door to an entirely new way of thinking about energy and matter. My work on the photoelectric effect won me the Nobel Prize. So perhaps the universe cares less about fame than about paying the bills.

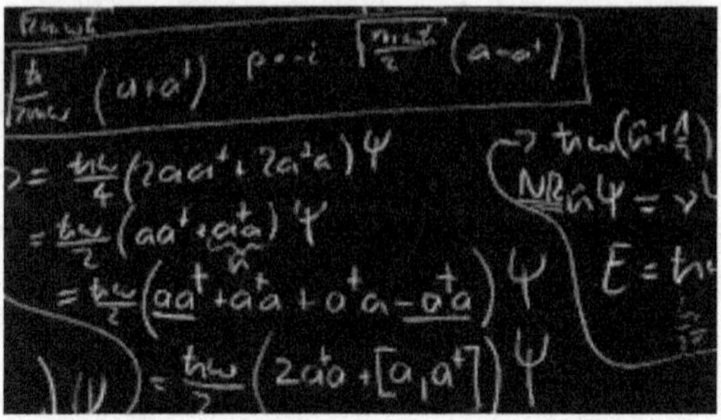

Q: What role should scientists play in politics and global affairs?

Albert Einstein: Science gives us powerful tools, but whether they are used for good or evil depends on human choice. Scientists, therefore, cannot stand aside from politics. We have a responsibility to speak out when knowledge is misused, especially when it threatens peace and human dignity.

A scientist should not seek power, but he must not be silent when his discoveries place power in the hands of those who might destroy. Politics is far more difficult than physics. At least in physics, the universe obeys laws.

Q: How do you feel about the fact that your theories indirectly led to the development of nuclear weapons?

Albert Einstein: I have always regretted that my equation $E = mc^2$ helped others unleash such destructive power. I did not intend to create weapons of mass destruction. Science is neutral; it is humanity that decides whether knowledge serves life or death. I have spent much of my later life advocating for peace and the responsible use of science. I discovered how to split the atom, but I never thought I'd be invited to the party of destruction. Some equations are better left in chalk, not in bombs.

Q: Do you believe science can make the world more peaceful, or is peace more a question of human character?

Albert Einstein: Science can give us tools, knowledge, and understanding, but it cannot guarantee peace. Peace depends on human character, on empathy, wisdom, and moral courage. Without these, science may even increase the power to harm. True progress requires both intellect and conscience.

Science can invent the airplane, the telescope, and the atom bomb — but only humanity can decide whether to use them wisely.

Q: You left Germany during a time of rising extremism. How did that experience shape your view of humanity?

Albert Einstein: It taught me how fragile civilization can be. A society rich in art and science descended into darkness almost overnight. Knowledge alone is not enough; we must also cultivate wisdom and compassion. It was easier to move across the Atlantic than to move prejudice out of the human heart.

Q: What does happiness mean to you?

Albert Einstein: A modest life, with simple pleasures — a good walk, a piece of music, a quiet thought — brings more peace than the restless pursuit of wealth or fame. A calm and modest life beats a life of endless champagne.

Q: You were known to play the violin often. How does music connect to science for you?

Albert Einstein: Music is not separate from science — both seek patterns, harmony, and beauty. When I play the violin, I feel the same curiosity and joy as when I explore the laws of the universe. Music helps me think, and sometimes the solutions to difficult problems come to me through melodies and rhythms rather than equations. If I cannot solve a problem with mathematics, at least I can play it on the violin until it makes sense.

Q: You had intimate relationships with many women. How did it impact your scientific work?

Albert Einstein: My personal life was sometimes complicated, and yes, I had relationships that were both joyful and challenging. I would not claim that they directly created my scientific ideas, but human emotions — love, curiosity, frustration — all feed the mind in subtle ways. Life experiences, including relationships, can inspire reflection and creativity, even in science. Romance may not give you equations, but it certainly gives you material to ponder while scribbling them.

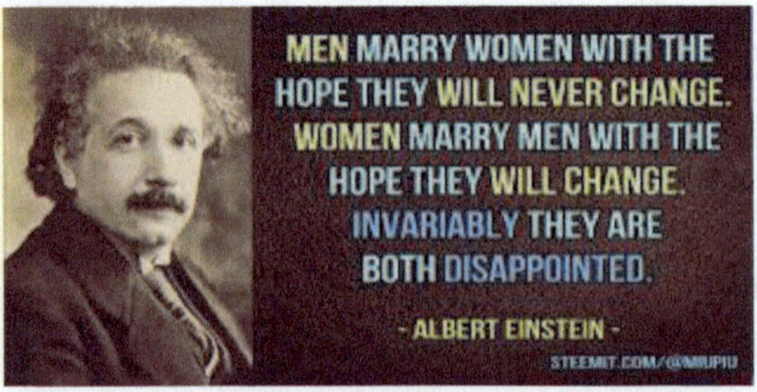

Q: Charlie Chaplin invited your wife and you in 1933 to the Hollywood premiere of his movie City Lights. What is your opinion about Charlie Chaplin?

Albert Einstein: Charlie Chaplin is a remarkable artist. His films combine humor with a deep understanding of human nature. I admire how he can make people laugh while also reflecting on social injustices. Art like his reminds us that imagination and empathy are as important as reason and science in understanding the world.

Q: When you met Charlie Chaplin, you had an interesting conversation. Please, tell us your comments about each other.

Albert Einstein: When I met Charlie Chaplin, I told him, "What I admire most about your art is its universality. You do not say a word, and yet the world understands you." It's true, replied Chaplin, "But your fame is even greater. The world admires you, when no one understands you."

Q: If you could give one piece of advice to young people who want to change the world, what would it be?

Albert Einstein: Do not fear, be different. The world changes not by those who follow the crowd, but by those who follow curiosity and conscience. Imagination will get you everywhere.

Closing Statement (Narrator)

As the conversation draws to a close, Einstein emerges not only as the architect of modern physics but as a moral voice for humanity. His ideas reshaped science, yet his wisdom extended far beyond the laboratory—into questions of war and peace, truth and responsibility, love and creativity. Though he left this world in 1955, his insights continue to guide us, reminding us that imagination is as vital as knowledge, and that science must serve humanity, not destroy it. Albert Einstein's voice remains a timeless beacon for those who seek to understand both the universe above us and the human spirit within us.

Nikola Tesla: Invention as a Way of Life

Introduction (Narrator)

Few minds in history have burned as brightly—or as mysteriously—as Nikola Tesla. Born in 1856 in a small village in the Austrian Empire, Tesla grew to become one of the greatest inventors and visionaries of all time. His genius gave the world alternating current, wireless communication, and countless innovations that laid the foundation for modern technology. Yet despite his brilliance, Tesla lived much of his life in solitude, often misunderstood, overshadowed, or dismissed by his contemporaries. In this interview, we step into the mind of a man who dreamed of harnessing the forces of nature, illuminating the world, and pushing the boundaries of human imagination.

Nikola Tesla (10 July 1856 – 7 January 1943) was a Serbian-American engineer, futurist, and inventor. He is known for his contributions to the design of the modern alternating current (AC) electricity supply system.

Q: What are your favorite inventions?

Nikola Tesla: The greatest achievement of my life was the Tesla coil, developed in 1891—a high-voltage, high-frequency transformer that enabled exploration of electricity and wireless energy transmission. This invention laid the foundation for my experiments with wireless power and radio. I also developed the complete system for the generation, transmission, and utilization of alternating current (AC)—though I did not invent AC itself. This system included AC motors, transformers, and polyphase power systems. The motors I provided to the world through Westinghouse are perhaps my most valuable contribution to human progress. Additionally, I demonstrated wireless radio communication before Marconi, who, unfortunately, received much of the credit for many years. In 1893, I publicly demonstrated radio in St. Louis, but the world was simply not prepared—it was far ahead of its time.

Q: What did you think of Edison?

Nikola Tesla: I respected Edison's relentless energy and dedication. If Edison had a needle to find in a haystack, he would diligently examine straw after straw until he found it, like a bee. However, I believed his trial-and-error method was inefficient compared to the scientific and mathematical approach I favored. I was a sorry witness to such toil, knowing that a little theory and calculation would have saved him ninety percent of his labor. Edison was a man of tremendous energy and persistence, with no hobbies or amusements, living in utter disregard of basic hygiene. He accomplished much, but at an enormous waste of energy. His work was empirical, lacking a deep understanding of nature's laws.

Q: Could you elaborate on the philosophical and technical differences with Edison?

Nikola Tesla: I championed alternating current (AC) for its efficiency in long-distance electricity transmission, while Edison supported direct current (DC). Our disagreement became known as the "War of the Currents." Edison even launched smear campaigns against AC, including public electrocutions of animals to portray it as dangerous. I saw his opposition as both misguided and commercially narrow-minded. Despite our rivalry, I believed Edison's approach belonged to a different era. His method was extremely inefficient, requiring immense effort unless aided by blind chance.

Q: Did you work with Edison on any projects?

Nikola Tesla: I worked briefly for Edison when I first arrived in the United States. I was promised a $50,000 bonus if I improved Edison's DC motors. After I succeeded, Edison allegedly dismissed the promise as a joke, saying, "You don't understand our American humor." I quit shortly thereafter. This experience left me feeling betrayed and deepened my disdain for Edison's character and business ethics.

Q: Were you jealous that Edison became wealthy while you did not?

Nikola Tesla: Jealous? No. I never measured success by money. Edison chose the path of enterprise and profit; I chose discovery and service to mankind. That path is rarely rewarded in dollars. Money does not represent the true value that men place upon it. My ideals led me to sacrifice all else, and they remain my true possession. Edison was pragmatic and played the business game well. I was perhaps too idealistic—giving away patents, turning down profits, and trusting too much in others' nobility. Let others chase wealth; I found joy in nature's secrets and the thrill of solving her puzzles—a richness beyond price.

Q: Did Edison steal any of your ideas?

Nikola Tesla: No, Edison did not steal my ideas. We worked differently and in different realms—he was a tinkerer and businessman; I was a scientist and dreamer. Our minds did not run on the same channels. That said, others did take my inventions and profited from them without acknowledgment. I do not accuse Edison directly, but the system he thrived in allowed such injustices. I gave the world the rotating magnetic field and alternating current; others patented around these concepts, marketed them, and reaped rewards. I did not concern myself with such matters—my goal was to uplift humanity, not compete for wealth or fame.

Q: Who was your favorite inventor?

Nikola Tesla: I greatly admired Michael Faraday, the English scientist who pioneered electromagnetism and electrochemistry. I also revered James Clerk Maxwell, who mathematically formalized Faraday's findings, and Galileo Galilei, a giant of scientific thought.

Q: May I ask you about your romantic relationships, if any?

Nikola Tesla: I never married. I do not believe many great inventions have been made by married men. I thought celibacy helped me concentrate on my scientific work, and romantic entanglements would only distract me. My mission to benefit humanity seemed incompatible with personal attachments. The life of an inventor is so intense and concentrated that one must forgo all human pleasures.

Q: Could you share your opinion about women?

Nikola Tesla: I deeply admired intelligent and accomplished women, but felt emotionally disconnected from them. Early in life, I idealized women. I had affection for a few, but no romantic involvement. The French actress Sarah Bernhardt reportedly charmed me, but nothing came of it. I had a warm, platonic friendship with Katherine Johnson, the wife of my friend Robert Underwood Johnson. I kept my distance from many women in society. I predict that as society evolves, women will eventually surpass men in intellectual pursuits.

Q: Could you explain your unusual romantic feeling for a white pigeon?

Nikola Tesla: I fed thousands of pigeons for years, but one—a beautiful, pure white female with light gray-tipped wings—was different. I only had to wish and call her, and she would fly to me. I loved that pigeon as a man loves a woman, and she loved me. As long as I had her, life had purpose. When she died, I felt a light had gone out of my life.

Q: Are you Serbian or Croatian?

Nikola Tesla: I always identified as Serbian. I was born in 1856 in Smiljan, then part of the Austrian Empire. My family was of Serbian Orthodox Christian heritage, and I strongly identified with Serbian culture and identity. Despite the geographic location of my birth, I considered myself Serbian.

Q: Did you enjoy traveling?

Nikola Tesla: I appreciated travel when it served my scientific and professional goals. I moved from my birthplace in the Austrian Empire to study in European cities like Graz and Prague, then emigrated to the United States to pursue my inventions and career. My work required me to be in places like New York, Colorado Springs, and Europe at times. While I traveled out of necessity and opportunity, I was not one to travel for leisure or adventure.

Q: Did you like art?

Nikola Tesla: I had a deep admiration for beauty and harmony, which influenced both my scientific work and personal tastes. I appreciated music—especially classical music—and had a refined sense of aesthetics. I often spoke about the importance of elegance and simplicity, not only in inventions but in nature and design, reflecting an artistic sensibility.

Q: Did you own a car, and did you know how to drive?

Nikola Tesla: I never owned or drove a car. I focused on my work and often relied on public transportation, walking, or horse-drawn carriages.

Q: Did you have enemies?

Nikola Tesla: I had some adversaries, mostly professional rivals rather than personal enemies. Edison campaigned actively against my AC system, sometimes using harsh tactics. Though largely professional rivalry, it was intense. I also faced conflicts with business partners and competitors who did not support my ideas or who undercut me financially, including difficulties with investors and industrialists like J.P. Morgan.

Q: What is your opinion about artificial intelligence?

Nikola Tesla: No answer, but **AI assistant answered instead**: Nikola Tesla, being a visionary inventor and scientist from the late 19th and early 20th centuries, might have been fascinated by the concept of artificial intelligence. Although AI as we know it didn't exist in his time, Tesla was deeply interested in electricity, machinery, and the potential of technology to improve human life. He might have seen AI as a natural extension of his work on automation and electrical engineering, possibly envisioning intelligent machines that could assist humanity in solving complex problems, advancing science, and enhancing daily life. Given his forward-thinking nature, he might also have pondered the ethical implications and the relationship between humans and intelligent machines.

Closing Statement (Narrator):

"As our conversation with Nikola Tesla draws to a close, one cannot help but sense both the brilliance and the solitude that defined his life. A man decades ahead of his time, Tesla dreamed of a world lit and powered by forces invisible to the eye, harnessed not for profit, but for the progress of

humanity. Though he often walked the path of a visionary without recognition in his lifetime, his ideas sparked revolutions in science and technology that continue to shape the modern world. His legacy reminds us that true genius is often misunderstood in its own age — yet it endures, waiting for the future to catch up."

Thomas Edison: An Imagined Conversation

Narrative Introduction

In the late 19th and early 20th centuries, few names carried as much weight in the world of invention and progress as Thomas Alva Edison. Known as "The Wizard of Menlo Park," Edison transformed modern life with inventions such as the phonograph, the motion picture camera, and most famously, the practical incandescent light bulb. With more than 1,000 patents to his name, his relentless spirit of experimentation and determination defined the age of innovation. Today, we imagine sitting across from Edison himself, peeling back the layers of the man who illuminated the world.

Thomas Alva Edison (February 11, 1847 – October 18, 1931) was an American inventor and businessman. He developed many devices in fields such as electric power generation, sound recording, and motion pictures. These inventions, which include the phonograph, the motion picture camera, and early versions of the electric light bulb, have had a widespread impact on the modern industrialized world. He was one of the first inventors to apply the principles of organized science and teamwork to the process of invention, working with many researchers and employees. He established the first industrial research laboratory.

Q*:* Of all your inventions, which one do you personally feel had the greatest impact on humanity?

Edison: That is not an easy choice, for each invention served its own purpose in shaping the world. The phonograph gave humanity a way to preserve its voice and music across generations. The motion picture allowed us to capture life itself and share stories beyond the limits of a stage. But if I must choose one, I would say the system of electric light and power had the greatest impact. The incandescent lamp was only a small part of it; the true achievement was creating an entire system of generation and distribution. It brought light to homes, extended the hours of productivity, and gave rise to modern cities that never sleep. Electricity became the lifeblood of progress, powering industry, communication, and medicine. To me, that was not merely an invention — it was the foundation upon which the future was built.

Q: You had little formal education as a child. How did that shape your approach to learning and invention??

Thomas Edison: Well, it's true I didn't have much formal schooling—just a few months, really. But I never thought of that as a handicap. If anything, it freed me from the rigid ways of thinking that schools sometimes impose. My mother was my great teacher. She believed in me and encouraged me to read everything I could get my hands on—history, science, and philosophy. I became a voracious reader, and that gave me a foundation broader than any classroom could have offered.

More importantly, I learned by doing. Every experiment, whether it failed or succeeded, taught me something. Formal education may teach you facts, but self-education and experimentation teach you how to think. That shaped my approach to invention: never be afraid to try, to fail, and to learn again. To me, curiosity was the best school I ever had.

Q: Your rivalry with Nikola Tesla has become legendary. How would you describe your relationship with him in reality?

Edison: Tesla is a brilliant man, no question about it. He had a remarkable mind for theory and a gift for envisioning things far ahead of his time. Our so-called rivalry has been turned into legend, but in truth, it was less personal and more a matter of business and philosophy. I was practical to the core — I believed in building systems that worked immediately and could be scaled to serve millions. Tesla, on the other hand, was a visionary who often leaped into ideas that the world was not yet ready to adopt.

We disagreed most fiercely on the question of electricity — my direct current versus his alternating current. That "War of Currents", as people call it, was really a battle of competing systems, investors, and companies. It wasn't so much about him and me personally, though it has been dramatized that way.

I respected Tesla's brilliance, even if I thought some of his ideas were impractical. And though we may not have been friends, I can say without hesitation that his contributions pushed the field forward. In the end, both of us helped electrify the world — each in our own way.

Q: Could you elaborate on the philosophical and technical differences with Nikola Tesla?

Edison: Certainly. At the heart of it, Tesla and I approached invention from two very different philosophies. I was an empiricist. I believed in relentless trial and error — testing thousands of filaments until I found one that would last, building systems piece by piece until they could serve people in their homes and businesses. My philosophy was: success comes from perspiration, persistence, and solving practical problems one at a time.

Tesla, by contrast, was a visionary theoretician. He could imagine entire systems in his mind, often without even sketching them out. Where I would fill notebooks with experiments, Tesla claimed he could design, refine, and test a machine entirely in his imagination before ever building it. That was a gift, but it also meant he sometimes leaped toward ideas that were decades ahead of their time — wireless power, global communication towers, or machines that the world had no infrastructure to support.

Technically, our most famous disagreement was over electricity. I advocated direct current (DC) because it was stable, safe at short distances, and immediately practical for lighting homes and businesses in cities. Tesla championed alternating current (AC), which could be transmitted efficiently over long distances with transformers. He was right about its scalability, and in time, AC became the dominant system for power distribution. But my focus was always on what could be built, installed, and monetized *today*, while his was on what might be possible *tomorrow*.

So, our differences were not only technical — AC versus DC — but philosophical: the pragmatist versus the dreamer, the builder versus the visionary. Both approaches mattered, for without Tesla's imagination and without my insistence on practical systems, electrification might not have advanced as quickly as it did.

Q: Did you work with Nikola Tesla on any projects?

Edison: Yes, briefly. When Tesla first came to America in 1884, he worked for my company — the Edison Machine Works in New York. He was fresh from Europe, full of energy and ideas. I assigned him to improve our direct current systems and motors. For a time, he worked diligently on those projects, and I saw his talent firsthand.

But our working relationship did not last long. Tesla wanted to pursue alternating current, and I was firmly committed to direct current at the time. That difference in direction created friction. There is a story often told — that I promised him a large reward for redesigning one of my machines and then dismissed it as a joke. The truth is more complicated, but misunderstandings like that soured his view of me.

So yes, we did collaborate for a short while. But Tesla was never one to remain under another man's vision. He had his own path to follow, and in time he struck out independently to pursue alternating current and other inventions. While we didn't remain partners, his early time at my company certainly gave him exposure to the practical side of the electric business in America.

Q: Some say you were not only an inventor but also a businessman. How did you balance creativity with commercial success?

Edison: To me, invention and business were never separate—they were two sides of the same coin. An idea in a notebook, no matter how clever, is useless if it cannot be brought into the hands of people who need it. I always believed that an invention isn't truly complete until it is practical, affordable, and available to the public.

That is why I built laboratories, not just workshops. I surrounded myself with talented men, we experimented tirelessly, and we tested not only for brilliance but for durability and cost. The phonograph, the light bulb, the electric system—none of these would have changed the world if they had remained expensive curiosities.

Yes, I was a businessman. I formed companies, I defended my patents, I negotiated with financiers. But I never saw profit as the enemy of creativity. Profit was proof that an invention had found its place in society. It meant we had solved not only a technical puzzle but a human need. I didn't. I married them. Invention gave me ideas; business gave them life.

Q: You also worked with George Westinghouse during the "War of Currents." Looking back, do you believe direct current (DC) could have ever won over alternating current (AC)?

Edison: At the time, I was convinced that direct current was the safer and more practical system. For city blocks and neighborhoods, DC did exactly what it was meant to do — provide steady, reliable light and power without the complications of high voltages. I built entire systems around it, from generators to distribution lines to lamps.

But the reality is that America was a vast country, and the world needed power to travel miles, not just blocks. Alternating current, championed by Tesla and driven into the marketplace by George Westinghouse, was far better suited for long-distance transmission. Once transformers made it possible to step the voltage up and down safely, AC's advantage became undeniable.

Could DC have won? Perhaps in small, dense cities, or if storage and conversion technologies had advanced more quickly. In fact, you still see DC in certain applications today — in batteries, in electronics, in rail systems. But for electrifying nations, AC was the logical victor.

If I regret anything, it's not the fight itself — competition pushes progress — but the excesses of that battle. The demonstrations, the campaigns to discredit AC... history remembers those more harshly than I ever intended. In truth, the world needed both approaches: my DC systems to light the first homes, and Tesla and Westinghouse's AC to carry power across continents. Together, they built the electric age.

Q: Who was your favorite inventor?

Edison: I had great admiration for many inventors who came before me. Michael Faraday, for instance, was one of my heroes. He had little formal education, much like me, yet his experiments with electricity and magnetism laid the very foundation for everything we built later. Without Faraday's discoveries, there would be no generators, no motors, no modern electrical industry.

I also respected Sir Isaac Newton, though he was more a scientist than an inventor. His relentless curiosity and systematic approach to uncovering natural laws inspired my own belief that nature's secrets could be unlocked through persistence.

But if I had to choose a favorite inventor in the practical sense, I would say Faraday. He was a true pioneer, a man who proved that knowledge and discovery did not belong solely to the privileged or highly educated. He showed that careful observation, experimentation, and imagination could change the world — and that's a lesson I carried with me in my own work.

Q: May I ask you about your romantic relationships, if any?

Edison: Well, my life was hardly the stuff of romance novels. I married twice — first to Mary Stilwell in 1871, when I was 24. She was a kind young woman who worked at my Newark shop, and we had three children together. Mary died tragically young in 1884.

Two years later, I married Mina Miller, the daughter of Lewis Miller, a prominent inventor and industrialist. Mina was not only my wife but also my anchor. She brought stability and refinement into my life at a time when I was consumed by work. We had three children together as well.

I was never a man for public displays of affection or scandalous stories. My laboratory was my great passion, but I was also deeply devoted to my family. Mina especially helped turn our home in West Orange into a true retreat — a place where, despite my long hours, I could find warmth and support.

So, while I may be remembered for electricity and phonographs, in private life I was simply a husband and father, grateful for the women who stood beside me while I chased my ideas.

Q: Could you share your opinion about women?

Edison: I always had respect for women, though I'll admit I came from an age when society often underestimated their potential. In my own life, women played a central role. My first wife, Mary, stood by me when I was still struggling to build my laboratories. My second wife, Mina, gave me a household of stability and culture that allowed me to focus on my work. She was educated, strong-willed, and managed the family in ways I never could have.

I observed that women possess qualities that men sometimes lack — patience, intuition, and a gift for understanding people. In truth, I believe invention is not just about machines and wires; it is also about seeing the world with empathy, and women often bring that perspective more naturally.

Had society afforded women the same opportunities as men in my day, I believe we would have seen many more inventors and scientists among them. I valued the contributions of women in education, the arts, and even in business. Though my own work was carried out largely with men in my laboratories, I never doubted that women were just as capable if given the chance.

Q: If you lived in today's world of artificial intelligence and digital technology, what would you focus on inventing?

Edison: The same principle that guided me in the 19th century would guide me now: identify what humanity needs most and then find a practical way to deliver it. In my day, that meant light, power, sound, and motion. Today, I see artificial intelligence, sensors, and digital networks as the new electricity — a force that can touch every corner of life if harnessed wisely.

I would focus on inventions that make these technologies *useful, understandable, and ethical* to ordinary people. Perhaps AI systems that explain their reasoning as clearly as a human teacher, medical devices that use machine learning to diagnose disease early, just as the stethoscope once transformed medicine, or new forms of energy storage and distribution to power an interconnected world without polluting it.

Another area would be education. When I was young, I had little formal schooling and had to teach myself. If I had today's tools, I would build learning systems — interactive, intelligent, affordable — so that every child could experiment and discover as freely as I did.

In short, I would do with AI and digital technology what I tried to do with electricity: turn a mysterious force into a *practical system* that improves daily life for millions, not just for a privileged few.

Q: Did you own a car, and did you know how to drive?

Edison: Yes, I owned several automobiles over the years. In fact, I was fascinated by them from the start — they were a perfect example of how invention could reshape daily life, just as electricity had. My good friend Henry Ford often brought me his latest models to inspect, and we spent countless hours discussing engines, fuels, and storage batteries.

I personally favored electric cars in the early years. Around 1901, I developed my own nickel–iron storage battery, which I hoped would power a new generation of electric vehicles. I owned a number of early electrics, and Mina and the children enjoyed riding in them.

As for driving — I did know how, though I was never as devoted to it as Henry Ford. I was more comfortable in the laboratory than behind the wheel. In truth, Mina and the chauffeur often drove me where I needed to go. But I kept a keen interest in technology, and I firmly believed the automobile, like electricity, would change how people lived and worked.

I suppose you could say I helped shape the car from the workshop, even if I wasn't always the one steering it down the road.

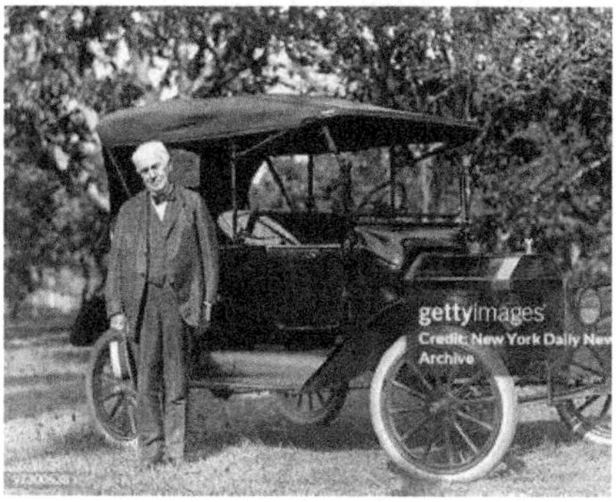

Q: Did you have enemies?

Edison: Enemies is a strong word. I certainly had rivals, critics, and competitors — some fierce ones. In the world of invention and industry, when money and progress are at stake, conflict is almost inevitable.

Nikola Tesla, of course, is often cast as my great adversary, though I've said before it was more a clash of ideas than of personalities. George Westinghouse was another competitor, particularly during the so-called 'War of Currents.' That rivalry was very public and sometimes bitter, but it was driven as much by investors and markets as by the men themselves.

I also had battles over patents — phonographs, motion picture equipment, and electrical systems. Many inventors and companies accused me of being too aggressive in defending my claims. Some even called me a monopolist. I admit I fought hard in the courts and the marketplace because I believed in protecting what my team and I had built.

Were these men enemies? I'd say they were opponents in the arena of invention. But I didn't carry hatred for them. Competition spurred me to work harder, to refine my ideas, and to keep pushing forward. If I had no rivals, perhaps I would have been less relentless. So, in a strange way, I owe as much to those who opposed me as to those who supported me.

Q: What advice would you give to young innovators and dreamers who want to change the world as you did?

Edison: The first thing I would tell them is this: ideas are cheap, but execution is priceless. Many people dream, but few are willing to put in relentless work to turn those dreams into something real. If you want to change the world, be prepared for long nights, failed experiments, and more discouragement than applause — at least in the beginning.

Second, never be afraid of failure. I failed thousands of times before producing a single working light bulb. But each failure taught me something. Persistence is what separates the dreamer from the inventor.

Third, focus on usefulness. An invention that cannot be applied, that does not solve a problem or make life better, is only a curiosity. Ask yourself constantly: How will this help people? How will it improve their lives?'

Lastly, surround yourself with good people. I did not invent myself alone. I had teams of machinists, chemists, draftsmen, and thinkers who helped me bring ideas to life. A great innovator is not only a thinker but also a leader who can inspire others to build together.

So my advice is simple: work harder than anyone else, learn from your failures, aim always for usefulness, and remember that even the greatest inventions are born from collaboration. Do that, and you too can leave something behind that lasts.

Closing Statement

As our conversation with Thomas Edison draws to a close, we are reminded that invention is rarely a stroke of luck; it is persistence, curiosity, and the courage to fail a thousand times in pursuit of one success. Edison once said, "Genius is one percent inspiration and ninety-nine percent perspiration," and his life's work stands as a testament to that belief. His innovations not only shaped the modern world but also inspired future generations of scientists, engineers, and dreamers to push beyond the known. In listening to Edison's words, we understand that the true legacy of an inventor is not just in the patents he leaves behind, but in the light he brings humanity's path forward.

Leonardo Da Vinci: An Imagined Interview

Narrator's Introduction

In the vibrant heart of the Italian Renaissance, one figure stood as a universal genius whose ideas, sketches, and visions stretched centuries beyond his time. Painter, engineer, anatomist, architect, and dreamer—Leonardo da Vinci embodied the limitless curiosity of the human spirit. From the ethereal smile of the Mona Lisa *to the intricate designs of flying machines, his mind roamed freely between art and science, searching for unity in all things. Today, we imagine a rare opportunity: to sit down with Leonardo himself, and to ask him about his life, his inspirations, and his vision for humanity.*

Leonardo di ser Piero da Vinci (15 April 1452 – 2 May 1519) was an Italian polymath of the High Renaissance who was active as a painter, draughtsman, engineer, scientist, theorist, sculptor, and architect. While his fame initially rested on his achievements as a painter, he has also become known for his notebooks, in which he made drawings and notes on a variety of subjects, including anatomy, astronomy, botany, cartography, painting, and paleontology. Leonardo is widely regarded as a genius who epitomized the Renaissance humanist ideal, and his collected works constitute a contribution to later generations of artists matched only by that of his younger contemporary, Michelangelo.

Q: Maestro Leonardo, you were born the son of a notary and a peasant woman in Vinci. How did your humble beginnings shape your outlook on life and your endless curiosity?

Leonardo da Vinci: I learned early that nature is the true teacher. I did not possess the schooling of the scholars, yet the rivers, the stones, the flight of birds, they became my books. What I lacked in Latin, I made up for in observation.

Q: You painted some of the most iconic works in history, such as *The Last Supper* and the *Mona Lisa*. Which of your paintings holds the deepest meaning for you?

Leonardo da Vinci: Each painting was an experiment, a search for truth. The *Last Supper* was about capturing a single moment of emotion—betrayal, devotion, uncertainty—all at once. The *Mona Lisa* was about the mystery of the human soul. But honestly, my favorite works are the ones I never finished. They live in my mind, always a little out of reach.

Q: Leonardo, many wonder—did you ever personally meet the woman known as the *Mona Lisa*? And what is your opinion of her?

Leonardo da Vinci: Yes, I met her. She was Lisa Gherardini, a Florentine woman, wife to Francesco del Giocondo. Yet what you see in the portrait is not simply her—it is something beyond her. She sat before me patiently, but as I painted, she transformed. Her face became a mirror of all women, of all humanity—serenity wrapped in mystery.

That smile… it is both real and imagined. Real, because it belonged to her. Imagined, because I shaped it to be eternal, to hold something that time cannot erase. My opinion of her? She was gentle, elegant, and kind. But the truth is, the *Mona Lisa* is not just Lisa—it is the marriage of her soul and my vision. She became my attempt to paint not a woman, but the very essence of life's quiet mystery.

Q: Leonardo, when you painted Lisa Gherardini — the woman we now call the *Mona Lisa* — did you ever feel love for her personally?

Leonardo da Vinci: Love? Not in the way poets write or the way lovers feel it. My feelings for Lisa were of admiration, not possession. She embodied a quiet grace, a stillness that fascinated me. When she sat before me, I did not see her only as a wife or a model but as a window into something universal.

If there was love, it was the love an artist feels for a subject — the devotion to understand, to reveal, to preserve. She inspired me not to desire, but wonder. Through her, I tried to paint not just her likeness, but the soul of humanity itself.

Q: Your notebooks are filled with designs—helicopters, tanks, even diving suits. Did you believe these inventions would ever exist in your lifetime?

Leonardo da Vinci: I knew the ideas were possible, but I also knew the world wasn't ready yet. My machines were screens meant for others to bring to life when the time was right.

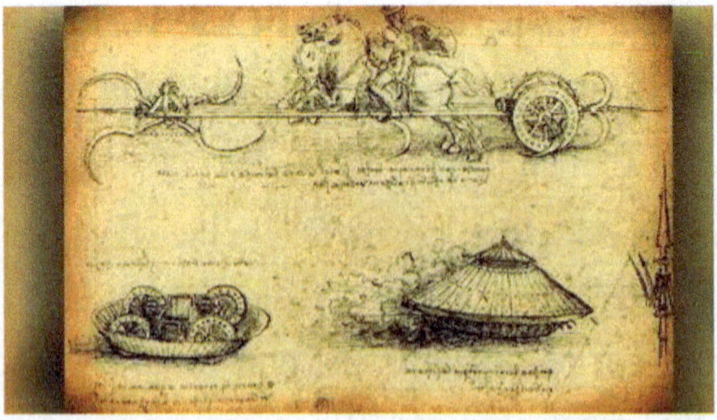

Q: You secretly studied human anatomy, even dissecting bodies. Why was that so important to you?

Leonardo da Vinci: Because to paint the human body, you have to understand it. Muscles, bones, veins—everything works in harmony. To me, the human body was the most perfect machine ever created, and I wanted to know how it worked from the inside out.

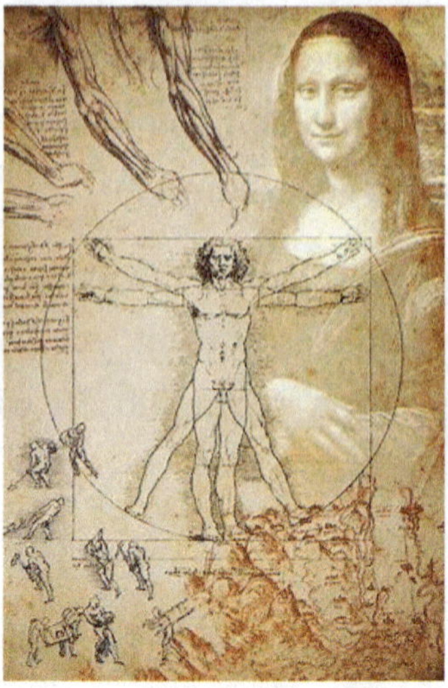

Q: You worked for powerful rulers, designing both beautiful works of art and weapons of war. Did you ever feel conflicted?

Leonardo da Vinci: Absolutely. Patronage kept me alive, but it didn't always feed my soul. I created war engines because my patrons demanded them, but what I truly loved was designing bridges, canals, and works that brought harmony and beauty.

Q: Leonardo, you've studied the human form, celebrated beauty, and painted some of the most iconic women in history. What is your opinion of women? Did you ever have a woman in your own life?

Leonardo da Vinci: Women, to me, are the embodiment of nature's greatest mystery. They give life, they nurture, they inspire — and yet they remain unfathomable, like the motion of water or the flight of birds. I observed women not as objects but as mirrors of the universe's harmony and complexity.

As for a woman of my own life… no. I never bound myself to marriage, nor to a single companion. My life was filled with work, study, and the pursuit of ideas. There were friendships, patronesses, muses, and models — and they were important to me. But my heart belonged not to one person, but to my curiosity and my art. Perhaps that was my fate: to love the mystery, rather than to possess it.

Q: If you could see our modern world—with airplanes, computers, and even space travel, how would you react?

Leonardo da Vinci: I would be amazed, of course. To see humans fly and walk on the moon—it's a dream come true. But I would also warn you: knowledge without wisdom can be dangerous. Invention should always serve humanity, not destroy it.

Narrator Closing:

As our imagined conversation with Leonardo da Vinci draws to a close, we are reminded of his timeless lesson: that curiosity has no limits, and that art and science, when joined, elevate humanity. His unfinished notebooks, sketches, and paintings are not marks of incompleteness but testaments to a restless spirit that never ceased to wonder. In Leonardo's world, every question led to another mystery, and every mystery to another possibility. His life whispers to us still: *"Saper vedere"—to know how to see.*

Steve Jobs: A Hypothetical Interview

Narrator Introduction

He was a pioneer who reshaped the way the world thinks, works, and dreams. Steve Jobs, the co-founder of Apple, stood at the crossroads of technology and artistry, insisting that innovation was not only about function, but about beauty and meaning. From the garage in Cupertino to the stages of Silicon Valley, his vision helped create devices that changed our daily lives and redefined entire industries. Today, we imagine a conversation with the man whose relentless pursuit of perfection made him both admired and controversial — but undeniably one of the greatest innovators of the modern era.

Steven Paul Jobs (February 24, 1955 – October 5, 2011) was an American businessman, inventor, and investor best known for co-founding the technology company Apple Inc. Jobs was also the founder of NeXT and chairman and majority shareholder of Pixar. He was a pioneer of the personal computer revolution of the 1970s and 1980s, along with his early business partner and fellow Apple co-founder Steve Wozniak.

Q: You were adopted and grew up in Silicon Valley. How did your upbringing shape your worldview?

Steve Jobs: Being adopted made me realize life is fragile but also chosen. I wasn't just born, I was picked. That gave me both freedom and responsibility.

Q: What did you take away from dropping out of Reed College — especially calligraphy classes?

Steve Jobs: Dropping out let me drop in. I went to the classes that inspired me. Calligraphy had no hope of practical use at the time — until the Macintosh. Suddenly, type and proportion mattered. That's the point: you can't connect the dots looking forward, only backward.

Q: Did you truly believe Apple would transform the world?

Steve Jobs: Absolutely. Not by making computers. By making *tools for the mind.* We always said — the computer is a bicycle for the mind.

Q: Why was integration — hardware and software — so essential to you?

Steve Jobs: Because people don't want to be system integrators. They want something that just works beautifully. That requires controlling the whole experience.

Q: You were famously demanding. Does greatness require being tough?

Steve Jobs: I wasn't there to be nice. I was there to build something insanely great. You push people because you see what they don't see in themselves.

Q: What makes a great entrepreneur?

Steve Jobs: Taste. Not in clothes or food. Taste in products. Knowing when something is truly great — and having the courage to kill everything else.

Q: Who were your mentors or role models when you were young, and how did they inspire you?

Steve Jobs: I didn't have many traditional mentors. But I was deeply influenced by people who thought differently. Edwin Land of Polaroid — he talked about the marriage of art and science, and I thought, "That's it." Bob Dylan and The Beatles reinvented themselves again. That gave me permission to do the same. And my father, Paul, taught me craftsmanship. He said: *Even the back of a cabinet, the part no one sees, should be perfect.* That stuck with me.

Q: How do you inspire a team to share your vision, especially when the path forward isn't clear?

Steve Jobs: You don't manage people with spreadsheets. You paint the picture of what's possible. You make them *feel* it. If the vision is powerful enough, people will walk through walls to get there. The path is never clear — that's the point. You're not leading them to where they already are. You're leading them to somewhere they've never been.

Q: What do you consider the qualities of a great entrepreneur?

Steve Jobs: Two things: vision and taste. Vision means seeing what doesn't exist yet but should. Taste means knowing the difference between good and truly great. And then you need sheer perseverance — because most people give up too early.

Q: How important is beauty in technology, beyond pure functionality?

Steve Jobs: Beauty matters. Functionality gets it into your life, but beauty makes you love it. And when you love something, you take care of it, you use it more, and it becomes part of you. That's the difference between a tool and a companion.

Q: How did being fired from Apple change you?

Steve Jobs: Best thing that ever happened. The heaviness of success was replaced by the lightness of being a beginner again. At NeXT and Pixar, I got to reinvent myself.

Q: Is failure necessary for innovation?

Steve Jobs: Yes. If you're not failing, you're not trying hard enough. The trick is to fail fast and learn.

Q: What lessons did you learn at NeXT and Pixar that influenced your second chapter at Apple?

Steve Jobs: At NeXT, I learned humility. You can build great technology and still fail if the market isn't ready. At Pixar, I learned the power of collaboration — film is the most collaborative art form there is. When I came back to Apple, I combined both: focus on timing, and build teams that create together, not in silos. That's what made the iMac, iPod, and iPhone possible.

NeXT taught me what not to do. Pixar taught me how to dream with others. Apple was where it all came together.

Q: Do you see yourself as an inventor, artist, or businessman?

Steve Jobs: None of those. I'm a storyteller. Technology was just my medium.

Q: How do you define success?

Steve Jobs: Did you create something that you're proud of? That's success. Money is secondary.

Q: What do you hope people will remember you for 50 years from now?

Steve Jobs: Not for me. For the work. That we built products with love, that mattered, and that made life a little better. I don't care if they remember me. I care if they remember the feeling our products gave them.

Q: If you could give one piece of advice to young innovators today, what would it be?

Steve Jobs: Don't chase money. Don't chase fame. Chase what makes your heartbeat faster. If you build something you love, the rest follows. Stay hungry. Stay foolish. That's not a slogan — that's survival.

Q: Did you have some close friends, and who were they?

Steve Jobs: I wasn't the easiest person to be friends with. I demanded a lot, sometimes too much. But I did have a few. Larry Ellison was one — we could talk about anything. At Pixar, John Lasseter became a real friend; we dreamed together. And a handful of people at Apple who shared the journey, not just the work. For me, friendship wasn't about hanging out. It was about building something meaningful side by side. My closest friends were the ones who built with me — not just the products, but the dreams.

Q: How did family shape your life?

Steve Jobs: For years, I was so obsessed with work that I thought family could wait. I was wrong. My wife Laurene grounded me. My kids taught me patience. You learn that building a family is harder than building a company — and more important. The most important creation in my life wasn't the Mac or the iPhone. It was my family.

Q: How do you see the future of technology shaping human life — are you more optimistic or cautious?

Steve Jobs: Both. Technology is like fire — it can cook your dinner or burn your house down. I'm optimistic because when technology meets creativity and humanity, it lifts us. I'm cautious because without values, it just amplifies the worst of us. The tools are neutral. What matters is the people who wield them.

Narrator Closing

As our dialogue with Steve Jobs concludes, we are reminded that his legacy is not merely in the products he helped build, but in the way he taught us to "think different." He proved that technology could inspire as much as it could serve, and that daring to follow one's passion can leave an indelible mark on the world. Though his journey ended too soon, the echoes of his vision remain in every pocket, on every desk, and in every dreamer who believes they can change the world.

Pablo Picasso: An Imagined Interview

Narrator's Introduction

He was a revolutionary of modern art; a man whose brushstrokes and vision shattered traditions and gave birth to entirely new ways of seeing the world. Pablo Picasso, born in Málaga, Spain, in 1881, was not only one of the most prolific painters of the 20th century, but also a sculptor, printmaker, ceramicist, and innovator whose influence touched every corner of artistic expression. From the blue melancholy of his early works to the bold experimentation of Cubism, Picasso redefined art itself, forever changing how we interpret form, color, and perspective. Today, we imagine a conversation with the man who once declared, "I do not seek, I find.

Pablo Diego José Francisco de Paula Juan Nepomuceno María de los Remedios Cipriano de la Santísima Trinidad Ruiz y Picasso (25 October 1881 – 8 April 1973) was a Spanish painter, sculptor, printmaker, ceramicist, and theatre designer who spent most of his adult life in France. Among his most famous works are the proto-Cubist *Les Demoiselles d'Avignon* (1907) and the anti-war painting *Guernica* (1937), a dramatic portrayal of the bombing of Guernica by German and Italian air forces during the Spanish Civil War.

Q: You showed extraordinary artistic talent as a child. Do you remember when you first realized painting would not just be a gift, but your destiny?

Picasso: I was born to paint. My first word was *piz*, short for *lápiz* — pencil. Some boys want to be soldiers; I wanted to hold the world in lines. My father taught me to paint doves; I decided I would paint everything else.

Q: **You moved through many artistic periods — the Blue Period, the Rose Period, and Cubism. What drives these transformations?**

Picasso: Life itself. My art has always been a mirror of my emotions, my struggles, my love, my politics. Blue for sorrow, rose for tenderness, Cubism for the chaos of modern life. I could never stay still, because to repeat myself would mean death.

Q: You co-created Cubism, a style that shattered perspective and reshaped modern art. What compelled you to break painting apart?

Picasso: Perspective was a prison. The world is not seen from one eye. I wanted to show the nose and the profile at once, the front and the side. Cubism was not destruction, it was liberation — a way to paint truth with lies.

Q: ***Guernica*** **became a universal symbol against war. Did you paint it more as an artist, as a Spaniard, or as a man enraged by human cruelty?**

Picasso: I painted it as a human being. The bombs fell on a town, and I answered with black, white, and grey. A scream on canvas. They asked me, 'Did you do this?' I told them, "No — you did."

Q: You were known for your intense relationships with women. How did they shape your art, and how did your art shape your relationships?

Picasso: Women are machines for suffering — and for inspiration. With each woman, I was reborn. Fernande was my youth, Olga my discipline, Dora my pain, Françoise my joy, Jacqueline my twilight. I painted them as I saw them — saints, demons, lovers, enemies. Sometimes all at once.

Q: Did you have a "true love," or was love for you inseparable from creation and destruction?

Picasso: My true love was painting. Women were rivers that fed it, but the canvas never betrayed me. Love destroys; art endures.

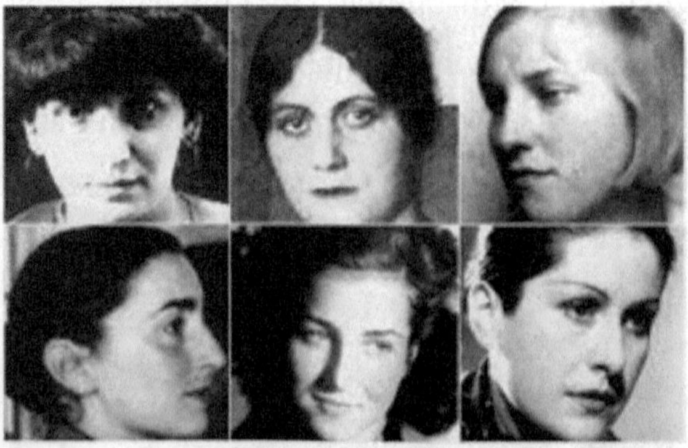

Q: You once said, *"Art is a lie that makes us realize truth."* What truth did you most hope people would find in your art?

Picasso: That nothing is fixed. A face, a body, a century — all can be broken and remade. The truth is not on the surface. You must shatter it first.

Q: What do you believe is the role of the artist?
Picasso: The artist is a political being, yes, but more than that, he is a seeker of truth. We do not decorate life; we expose it, we challenge it, we transform it. My role was to open eyes — to make people see the world anew.

Q: If you could choose how future generations remember you — painter, sculptor, revolutionary, genius, or tyrant — which word would you keep?

Picasso: I want to be remembered as a painter who never stopped being hungry. Genius and tyrant are words for critics. I prefer *living*.

Narrator Closing

As our imagined conversation with Pablo Picasso ends, we are left with the portrait of a man who refused to be bound by convention. He lived as he painted — fiercely, endlessly, and without apology. His legacy is not only in the canvases hanging in the world's great museums, but in the very way we understand art itself: as a force that provokes, heals, and reveals. Picasso's voice reminds us that creativity is not a gift for the few, but a power within us all — to see differently, and to live more fully.

INTERVIEW WITH
PABLO PICASSO

Q I was born to paint… Some boys want to be soldiers, I wanted to to hold the world in lines. My father taught me to paint doves; I decided I would paint everything else.

Q I painted it as a human being. The bombs fell on a town, and I answered with black, white, and grey. A scream on canvas.

Q Women are machines for suffering—and for inspiration… I painted them as I saw them—saints, demons, lovers, enemies. Sometimes all at once.

Q Art is a lie that makes us realize truth… A face, a body, a century—all can be broken and remade. The truth is not in the surface. You must shatter it first.

Q I want to be remembered as a painter who never stopped being hungry. Genius, tyrant—those are words for critics. I prefer: alive.

John Lennon: Imagine an Interview

Narrator's Introduction:

John Lennon was never just a musician. He was a rebel, a poet, a dreamer, and a provocateur who used his voice and guitar to question authority, challenge norms, and imagine new worlds. From the streets of Liverpool to the heights of Beatlemania, from the peace protests with Yoko Ono to his quieter years as a father, Lennon lived a life of contradictions: tenderness and rage, fame and isolation, wit and vulnerability. In this conversation, his words reveal the man behind the myth — sharp, restless, and searching for truth.

John Winston Ono Lennon (born **John Winston Lennon**; 9 October 1940 – 8 December 1980) was an

English musician and activist. He gained global fame as the founder, co-lead vocalist, and rhythm guitarist of the Beatles. Lennon's songwriting partnership with Paul McCartney remains the most successful in history.

Q: What was your childhood in Liverpool like, and how did it shape your music?

Lennon: Liverpool wasn't exactly a playground, you know. It was tough, gray, post-war Britain. My mum was here and gone, my dad wasn't around, and Aunt Mimi raised me. It gave me a sharp tongue and a soft heart I tried to hide. The roughness gave me rebellion; the music gave me escape.

Q: How did you first meet Paul McCartney, and what convinced you he was the right partner?

Lennon: Paul turned up at a church fête, strumming his guitar better than most lads I'd seen. He knew the chords, and I thought, 'Right, this one's dangerous — keep him close.' We balanced each other. I was the cynic, he was the romantic. Together, we made magic.

Q: When The Beatles became a worldwide phenomenon, did you feel liberated or trapped by fame?

Lennon: Both. The fame gave us everything — money, women, freedom to do what we wanted. But it also put us in a bloody cage. Everywhere we went, screaming mobs. I couldn't hear the music, couldn't breathe. Fame is like candy — sweet at first, then it rots your teeth.

Q: Looking back, which Beatles song do you feel captured your soul the most?

Lennon: Probably *Help!* People thought it was just another pop tune, but it was me crying out, saying, 'I'm drowning here.' That's the funny thing — you can hide the deepest truth in a catchy melody, and no one notices until years later.

Q: Your lyrics often blend vulnerability with rebellion. Where did that balance come from?

Lennon: Because I was both — soft and hard. I wanted love, but I also wanted to smash the system. One minute I'm writing 'Imagine,' the next I'm spitting at the establishment. That's life, isn't it? You're never just one thing.

Q: How did experimenting with new sounds and styles change your view of music as an art form?

Lennon: Music's just colors and shapes to me. We went from rock 'n' roll to sitars to tape loops and back again. Every experiment was a way of saying, 'We don't have to follow the bloody rules.' Art dies the moment you stop playing with it.

Q: Did you consider yourself more of a poet who set words to music, or a musician who used poetry?

Lennon: Neither really. I was a thief — I stole words from books, riffs from records, ideas from the air. I just put them together in a way that felt true. Poet, musician, whatever — I was just trying to make sense of myself.

Q: What inspired you to use music as a platform for peace and protest?

Lennon: Because politicians weren't doing it. Because wars kept happening. Because if you've got a guitar and a microphone, you've got a weapon louder than a gun. Singing 'Give Peace a Chance' wasn't just a song — it was an idea anyone could join in.

Q: During your "Bed-Ins for Peace" with Yoko, did you believe art could really change politics?

Lennon: Maybe not politics, but people. And people change politics. The bed-ins were daft, yeah — two scruffy artists in pajamas, talking peace to the press. But it made people laugh, think, maybe dream for a moment. That's all art must do — open a crack.

Q: How did your partnership with Yoko Ono influence your creativity and worldview?

Lennon: Yoko blew the walls off my mind. She showed me that art could be anything — a note, a silence, a scream, a sky. People blamed her for breaking up the Beatles, but the truth is, I was ready to move on. She didn't break us — she freed me.

Q: Looking back, what was the most misunderstood part of your relationship with her?

Lennon: That she controlled me. People painted me as some puppet. Rubbish. We were equal. Sometimes she led, sometimes I did. That's what partnership is. The world wasn't ready for a woman standing next to a Beatle with her own ideas.

Q: Do you feel that fatherhood changed you more than fame ever did?

Lennon: Absolutely. Fame gives you ego, kids give you humility. With Sean, I became a househusband, baking bread, changing nappies. That was more real than all the stadiums in the world. Being a father grounded me in a way the Beatles never could.

Q: If you could speak to your younger self at the start of Beatlemania, what advice would you give?

Lennon: Slow down. Don't chase every height. And for God's sake, don't be so cruel to the people who love you. I had a sharp tongue and hurt people I cared for. I'd tell young John: the world will love you, but don't forget to love back.

Q: Do you think The Beatles' breakup was inevitable, or could it have been avoided?

Lennon: It was inevitable. We grew up, we grew apart. Four lads from Liverpool couldn't stay freezing in time forever. Better to end it and let the music stand than drag it on until it dies of boredom.

Q: What do you want people to remember most: your music, your activism, or simply you as a man searching for truth?

Lennon: Remember me as a dreamer who tried. The songs will outlive me, the peace marches will fade, but if one person hears 'Imagine' and believes a better world is possible, then I did my job.

Narrator's Closing Statement:

John Lennon's life was cut short, but his voice still echoes — in music, in protest, in the simple dream of peace. He challenged the world to see differently, to love more fiercely, and to imagine what could be. In the end, Lennon's truth was not perfection but honesty. He was flawed, he was brilliant, and he was human. As he once sang: "You may say I'm a dreamer, but I'm not the only one."

Elizabeth Taylor: An Interview That Never Happened

Introduction (Narrator)

Elizabeth Taylor, often called "Liz," was one of the most iconic and glamorous actresses of the 20th century. With her striking violet eyes, timeless beauty, and undeniable talent, she captivated audiences across decades. From her breakthrough as a child star in National Velvet *to her Academy Award–winning performances in* Butterfield 8 *and* Who's Afraid of Virginia Woolf? *Taylor became a legend both on and off the screen. Beyond Hollywood, she was also admired for her resilience, her eight marriages, and her passionate activism in raising awareness for HIV/AIDS at a time when few dared to speak about it. Today, we revisited her extraordinary journey through an imagined conversation.*

Dame Elizabeth Rosemond Taylor (February 27, 1932 – March 23, 2011) was an English and American actress. She began her career as a child actress in the early 1940s and was one of the most popular stars of classical Hollywood cinema in the 1950s. She then became the world's highest-paid movie star in the 1960s, remaining a well-known public figure for the rest of her life. In 1999, the American Film Institute ranked her seventh on its greatest female screen legends list.

Q: You began acting at a very young age. How did growing up in the spotlight shape your personality and outlook on life?

Elizabeth Taylor: It forced me to grow up quickly. I never really had a normal childhood. From the time I was ten, people looked at me not as a little girl but as an actress, a commodity almost. But I also learned resilience. Hollywood can be cruel, and if you don't know who you are, it will tell you who to be. I decided early on I would never let it own me completely.

Q: *National Velvet* made you a household name. Did you realize at that time that you would become a Hollywood legend?

Elizabeth Taylor: Not at all. I was just a girl who loved horses. Acting was fun, but I didn't think of myself as a star.

I never thought in terms of legacy—honestly, I just wanted to do good work, live passionately, and be true to myself. The rest came as a byproduct.

Q: You worked with so many of Hollywood's greatest leading men. Which co-star do you remember most fondly, and why?

Elizabeth Taylor: Oh, there were so many! Montgomery Clift was one of my dearest friends—we had a soul connection, pure and deep. Working with him in *A Place in the Sun* was unforgettable. And of course, Richard—he wasn't just a co-star, he was part of my life. On screen, off screen, our chemistry was fire and storm all at once.

Q: What are your favorite movies?

Elizabeth Taylor: That's like asking a mother to choose her favorite child — impossible. But some films were dearer to me than others, for different reasons. ***National Velvet*** will always have a special place in my heart, because it was my beginning, and because I adored horses more than anything at that age. I was just a little girl, but it changed my life.

Then there's ***A Place in the Sun***. Working with Monty Clift was one of the greatest joys of my career. We were soulmates in many ways, and I think you can see that on screen. That film made me feel like a real actress for the first time."

"Of course, I can't forget ***Who's Afraid of Virginia Woolf?*** That nearly killed me — the emotional toll was immense — but it's the performance I am most proud of. To strip away glamour and bare the rawest parts of myself… it was terrifying and liberating at once.

People often mention ***Cleopatra***, and yes, it was spectacular — and infamous! But to me, that film was less about the costumes and sets, and more about what happened off-screen with Richard. It was the beginning of a love affair that defined so much of my life.

I don't have just one favorite. My films are like chapters of my life. Some were joyous, some painful, some transformative. Taken together, they are my stories.

Q: You were often called the most beautiful woman in the world. How did you personally handle the attention and scrutiny that came with that?

Elizabeth Taylor: To be honest, I never saw myself the way the world did. Beauty is fleeting, and it can be a trap. People expect perfection from you, and that's impossible. I never wanted to be remembered just for how I looked—I wanted to be known for my work, my courage, and my heart.

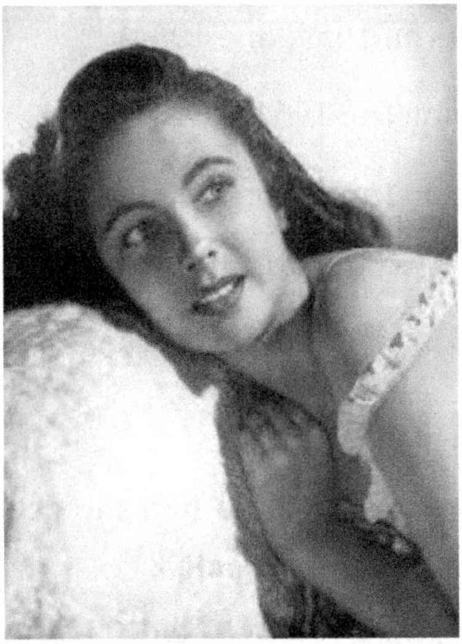

Q: You lived your love life very openly, marrying eight times to seven men. Do you believe the public ever truly understood your search for love and companionship?

Elizabeth Taylor: Probably not. People saw the headlines, the scandals, but they didn't see me—Elizabeth, the woman who simply wanted love, security, passion. I've always believed in marriage, in commitment. Perhaps I believed too much. But I'd rather be a woman who loved deeply and often than one who never dared at all.

Comment: Elizabeth Taylor was married 8 times:

Conrad Hilton Jr. (m. 1950; div. 1951)

Michael Wilding (m. 1952; div. 1957)

Mike Todd (m. 1957; died 1958)

Eddie Fisher (m. 1959; div. 1964)

Richard Burton (m. 1964; div. 1974)

Richard Burton (m. 1975; div. 1976)

John Warner (m. 1976; div. 1982)

Larry Fortensky (m. 1991; div. 1996)

Q: Which of your marriages had the deepest impact on you, personally and emotionally?

Elizabeth Taylor: Each marriage shaped me in some way, but Richard Burton—he was the great love of my life. We fought, we adored, we destroyed, and we healed each other in turns. With Richard, it was never quiet; it was opera. And even when it ended, it never truly ended.

Q: You converted to Judaism in 1959. Could you share what led you to that decision and how it shaped your beliefs?

Elizabeth Taylor: It wasn't something I did lightly or for show. I was raised in a Christian household, but as I grew older, I found myself drawn to Judaism's values — its emphasis on family, tradition, learning, and, most of all, on the sanctity of life and compassion for others. I studied for a year with Rabbi Max Nussbaum at Temple Israel of Hollywood before my conversion, and the more I learned, the more it felt like home to me spiritually.

People assumed at the time that I converted because of Eddie Fisher, but that wasn't true. My decision was deeply personal and preceded our marriage. Judaism gave me a sense of continuity and rootedness in a world that often felt chaotic. It also gave me a moral compass that I carried into my work and activism.

I've always believed that faith isn't about labels or exclusivity; it's about how you live your life — the kindness you show, the justice you fight for, the humility with which you walk. That's what Judaism taught me. It helped me become not just a better actress or a better wife, but, I hope, a better human being.

Q: You faced serious health issues throughout your life. How did you find the strength to keep returning to the screen and to your causes?

Elizabeth Taylor: I suppose I've always been stubborn. Pain never frightened me as much as giving up did. My battles with illness gave me empathy, and maybe that's why I was drawn to causes like HIV/AIDS. I knew what it felt like to be in pain, to be vulnerable, and I wanted to stand by those who were suffering.

Q: You were one of the first celebrities to speak openly about HIV/AIDS and established the Elizabeth Taylor AIDS Foundation. What motivated you to take on that cause so passionately?

Elizabeth Taylor: Because people were dying—and they were being ignored. My friends were dying. The stigma was so cruel, and I couldn't sit by. I used my fame, my platform, to speak when others were silent. That may be the proudest thing I've ever done in my life.

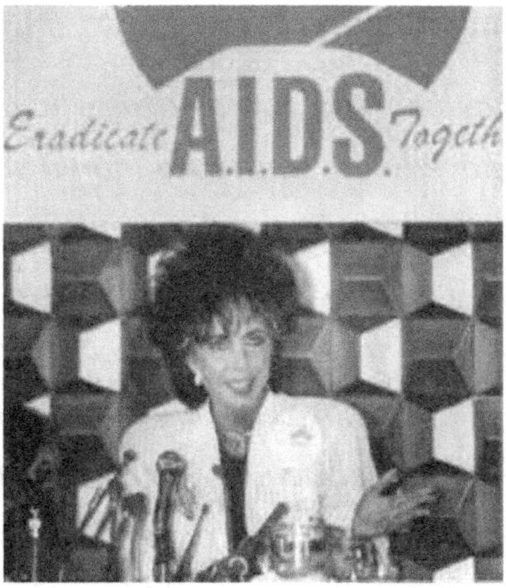

Q: Looking back, do you feel your greatest legacy lies in your films, your activism, or in the way you lived your life so boldly?

Elizabeth Taylor: Perhaps all of it, but if I had to choose—my activism. Movies fade, beauty fades, but compassion has a way of living on. If people remember me as someone who stood up for the vulnerable, then I'll rest easily.

Q: If you could share one piece of advice with women of future generations, what would it be?

Elizabeth Taylor: Never apologize for who you are. Love passionately, fight fiercely, and don't let anyone define you. Life is short, and the only sin is not to live it fully.

Closing Statement (Narrator)

Elizabeth Taylor's life was as dramatic and dazzling as the roles she played. She was more than a starlet—she was a woman of strength, vulnerability, and fierce compassion. Her beauty may have been timeless, but her true brilliance lay in her courage to love unapologetically, to fight for what mattered, and to remain authentic even when the world expected perfection. In the end, Elizabeth Taylor was not just Hollywood royalty; she was a force of nature who left behind a legacy of artistry, resilience, and humanity.

Marilyn Monroe: Beyond the Screen

Introduction (Narrator)

She was more than a movie star—she was an icon. Marilyn Monroe, born Norma Jeane Mortenson in 1926, rose from a troubled childhood to become the most recognizable face in Hollywood. With her platinum hair, breathless voice, and dazzling screen presence, she captivated the world. Yet behind the glamour was a woman of contradictions—vulnerable yet ambitious, adored yet lonely, playful yet profoundly intelligent. In this interview, Marilyn reflects on her life in the spotlight, her struggles with fame, and the dreams that defined her beyond the silver screen.

Marilyn Monroe was an American actress, comedienne, singer, and model. Monroe is of English, Irish, Scottish, and

Welsh descent. She was born in 1926 and died in 1962. She became one of the world's most enduring iconic figures and is remembered both for her winsome embodiment of the Hollywood sex symbol and her tragic personal and professional struggles within the film industry. Her life and death are still the subjects of much controversy and speculation.

Q: How many marriages did you have, and who was your favorite husband?

Marilyn Monroe: Well… let's see… I've been a bride three times. Once, when I was just a little girl, before I even knew who Marilyn Monroe was. Then to Joe… you know, the handsome Yankee Clipper. And then to Arthur, who made me feel very clever. One too soon, one too short, and one too serious.

As for a favorite… oh, I don't think a lady should play favorites with her ex-husbands. But… let's just say Joe always made me feel safe. He never really stopped caring about and I liked that very much.

Comment: Marilyn Monroe was married **three times**:

1. **James Dougherty** (1942–1946) – She married him at age 16, before she was famous.
2. **Joe DiMaggio** (1954–1955) – the baseball legend; their marriage lasted only 9 months, but he remained deeply devoted to her after the divorce.
3. **Arthur Miller** (1956–1961) – the playwright; their marriage lasted 5 years.

Q: What are your favorite movies?

Marilyn Monroe: I suppose most people would say Gentlemen Prefer Blondes—because I got to sing and dance and wear all those beautiful gowns. Diamonds really are a girl's best friend, you know. I think I liked Bus Stop best…

it was the first time I felt people saw me as more than a joke in a tight dress. I worked very hard on that role, and it mattered to me.

Q: You've worked with some of the biggest stars in Hollywood. Who were your favorite leading men?

Marilyn Monroe: Oh… that's a hard one. I've been very lucky. Clark Gable was just wonderful — he really was the King, and I felt so safe with him. I used to watch his films when I was a little girl, and then suddenly there I was acting with him. That was a dream come true.

Q: And what about Montgomery Clift? You seemed close during The Misfits.

Marilyn: "Monty was very special. He was delicate, like me, and he had such a real soul. We understood each other without words sometimes. I loved working with him because there was no acting — it just felt honest.

Q: Fans also remember your chemistry with Tony Curtis and Yves Montand.

Marilyn (smiling): "Tony made me laugh all the time — even when we were supposed to be serious. And Yves… well, let's just say he was very charming both on and off the set."

Q: So, if you had to pick just one?

Marilyn (playfully): Oh, a lady never picks just one. But in my heart… Clark Gable will always be my favorite.

Q: You sang for President John Kennedy on his birthday. Could you tell us about your relationship with JFK?

Marilyn Monroe: I was invited to sing for the President on his birthday. It was quite an honor — not every girl gets to serenade the most powerful man in the world. I just wanted to make it memorable… and maybe a little sexy too.

Jack was charming, no doubt about it. With him, you felt the whole room disappear. But men like him don't belong to one woman, and I was too lonely to only be someone's secret. Still, when I sang that night, it was just for him.

Q: Did you also know his brother, Robert Kennedy?

Marilyn Monroe: Yes, I knew Robert, too, mostly at public gatherings. He was very kind to me, very polite. But my real acquaintance was with the President — I admired his vision for the country. Robert was different from Jack — less of the showman, more serious, but still warm. With Bobby, I sometimes felt listened to, really listened to. But I also knew I could never truly belong in their world.

Q: Did you like partying?

Marilyn Monroe: Hollywood is always having parties… and sometimes I show up if they promise there's good champagne. People think I love big parties, but I really don't. I get nervous in crowds, and I never feel like I fit in. I prefer a quiet evening with a few close friends. Or just listening to music by myself.

Q: Did you have some favorite friends?

Marilyn Monroe: I think everyone's my friend when there's a camera around! But the real ones — they're the people who don't care if I'm Marilyn Monroe or just Norma Jeane. Those are the ones I treasure. I've had a few dear friends. Natasha Lytess, my drama coach, believed in me when I didn't. Also, Ella Fitzgerald, whom I helped get a booking at the Mocambo Club. And I always trusted my makeup man, Allan 'Whitey' Snyder — he was with me from the start.

Q: How important was money in your life?

Marilyn Monroe: I've always been afraid of not having enough. Even when I earned millions, I worried if it would last. Money isn't what makes life happy… but without it, it's hard to survive. Sometimes I think that fear has haunted me more than anything else. Money was never the goal. I wanted to be respected for my work, for my talent. But I did need it — to pay my bills, to feel independent, to protect myself from being taken advantage of. That was very important to me.

Q: In 1949, you posed in a nude calendar, which caused a scandal. What is your opinion on that?

Marilyn Monroe: Oh, yes, that little calendar! It certainly got people talking. Those photos… they were a part of my past. I was young, ambitious, and trying to survive in Hollywood. I don't think they define me as a person or as an actress. People loved to judge me for it, but I've always tried to move forward. But I wouldn't pose like that today… Well, maybe if the price were right!

Q: Your clothing has been shockingly revealing or even malfunctioning, such as when a shoulder strap of your dress snapped during a press conference. Could you comment on this?

Marilyn Monroe: Oh, that dress! Well… sometimes a girl just wants to make an entrance. And sometimes things don't cooperate! You have to laugh, don't you? I think the cameras enjoy it more than I do. I don't mind showing a little skin — it's part of the image, part of the performance. But wardrobe malfunctions… those are always embarrassing! I try to stay graceful, even when a strap breaks or a hem falls.

Q: Is it true that you refused to date Elvis Presley?

Marilyn Monroe: No answer.

AI assistant answered instead:

While I can't speak for Marilyn Monroe personally, historically, there's no confirmed evidence that she "refused to date Elvis Presley." The story seems to be more of a rumor circulated by tabloids. Marilyn and Elvis were both huge cultural icons in the 1950s. Elvis admired Marilyn and even mentioned her in interviews, but there's no verified account of them having serious romantic interaction. Some sources suggest that Marilyn might have been cautious around younger men or simply uninterested in a relationship with Elvis, who was about 8 years younger than her.

Closing Statement (Narrator)

As our conversation ends, we are left with a portrait of Marilyn not as the myth, but as the woman—complex, searching, and achingly human. Her life was a fragile balance between the persona the world demanded and the person she longed to be. Though her time was tragically short, Marilyn Monroe's presence endures, not only in her films but in the way she embodied the hopes, insecurities, and desires of an era. Decades later, her image still glows on the screen, reminding us that legends never truly fade—they live on in memory, in art, and in the longing of those who continue to be captivated by her light.

Frank Sinatra: The Soundtrack of a Century

Introduction (Narrator)

Few entertainers have defined an era as the way Frank Sinatra did. Known simply as "Ol' Blue Eyes," Sinatra's voice became the soundtrack of the 20th century—smooth, powerful, and unforgettable. He was more than a singer: an Academy Award–winning actor, a cultural icon, and a man whose life was as complex and colorful as the songs he sang. From the heights of his fame with the Rat Pack to his moments of personal struggle, Sinatra embodied both glamour and grit. In this rare conversation, we hear Sinatra reflect on his artistry, his influences, his relationships, and the price of living in the spotlight.

Francis Albert Sinatra (December 12, 1915 – May 14, 1998) was an American singer and actor. Nicknamed the "**Chairman of the Board**" and "**Ol' Blue Eyes**", he is regarded as one of the most popular entertainers of the 20th century. Sinatra is among the world's best-selling music artists, with an estimated 150 million record sales globally.

Q: You are known for your impeccable phrasing and timing. How did you develop your unique singing style, and what was the role of instinct versus training?

Sinatra: You know, kid, I never thought of myself as just a singer. I thought of myself as a storyteller. The words came first—the music was there to carry the meaning. I learned phrasing from Tommy Dorsey. I'd listen to that trombone of his, how he could hold a note like forever. I'd go home and practice breathing, swimming underwater, anything to stretch it out.

But instinct—that's the secret sauce. Training gave me the tools. Instinct told me when to bend the rules. Sometimes you sit right on the beat, sometimes you lean behind it, make the audience wait a half-second longer. That's what makes people lean in.

Q: You transitioned from music to film seamlessly. What did acting give you that singing didn't?

Sinatra: Movies gave me a whole new way to tell stories. On stage, it's just me, the mic, and the band. In film, you've got the camera, the lights, the script, other actors—all these moving parts. It made me dig deeper, find emotions I might not touch in a song. And let me tell you, winning that Oscar for *From Here to Eternity* wasn't just an award. That was a second chance at a career people thought was over.

Q: The Rat Pack became a cultural symbol of an era. Did you see it as friendship, performance, or something larger?

Sinatra: It started as friendship—me, Dean, Sammy, Joey, Peter—guys who liked each other's company. But it turned into a show, no doubt about it. Vegas lit up when we were together. We weren't just singing and joking; we were selling a kind of dream. Suits, cocktails, laughs, girls, the whole package. Maybe too much of a dream sometimes. But hell, it was fun while it lasted.

Q: They admired your confidence and style. Was there ever insecurity behind the swagger?

Sinatra: Every night before I went on stage, I was nervous. People don't believe that. But if you don't get butterflies, you don't care enough. The swagger—that was armor. Inside, I was just hoping I wouldn't let the song down, wouldn't let the people down. Confidence is half real, half act. You have got to sell it either way.

Q: You've been called both generous and difficult. Which side do you think defines you more?

Sinatra: I'm Italian, I'm from Hoboken—what do you expect? (laughs) I had a temper, sure, but I had a big heart too. If I liked you, I'd go through fire for you. If you crossed me, forget it. Which side defines me? Both. You can't split a man in half.

Q: You had friendships with presidents and mob figures alike. Did you feel caught between two worlds, or did you see no contradiction?

Sinatra: Look, I know a lot of people. Some wore suits on Pennsylvania Avenue, some wore suits in dark clubs in Jersey. People love to spin stories, but the truth is, I didn't need the mob to get work. I had talent, I had the band, I had the records. Did I know guys with shady reputations? Sure. Everybody did in those days. But I wasn't taking orders from anybody.

Q: There have always been rumors about mob ties. How do you respond to that?

Sinatra: Rumors sell papers. People like scandals. But I'll tell you something—I built my career on my voice, not on favors. The mob didn't teach me how to phrase lyrics. They didn't teach me how to hold an audience. That was all for me. Anyone who thinks otherwise doesn't know squatting about music.

Q: You had several high-profile marriages and relationships. How did your personal life affect your work?

Sinatra: Love was my fuel. Every time I fell in love—or fell apart—I sang better. Ava Gardner, for instance… she broke me in half, and out came some of the best records I ever made. Happiness didn't make me sing; heartbreak did. Maybe that's the curse of a singer.

Q: You were known for loyalty to your friends but also for feuds. Did you hold grudges?

Sinatra: You're damn right I did. If someone wronged me, I didn't forget it. But loyalty—loyalty meant everything to me. Sammy, Dean—those guys were like family. If you were in my circle, I'd bleed for you. If you crossed me? Well, I wasn't the forgiving type.

Q: You broke racial barriers by standing with Black performers like Sammy Davis Jr., even when it cost you. Why was that important to you?

Sinatra: Because it was the right thing to do. Sammy was my brother. I didn't care if a casino didn't want him at the table—I wasn't playing unless he was. I grew up in Hoboken; I knew what it was like to be pushed around. Nobody was going to push my friends around if I had a say in it.

Q: If you could relive one night—on stage, in the studio, or off the spotlight—what would it be, and why?

Sinatra: Vegas, Sands Hotel, '60s. Count Basie's band behind me, Quincy Jones conducting. Dean and Sammy cutting up in the wings. The room is electric, everybody's dressed in the nines, martinis clinking. I felt ten feet tall. If I could bottle that night, I'd never need another drink again.

Q: You were famously close to John F. Kennedy, but then suddenly cut off. What really happened there?

Sinatra: Jack was a friend. When he was running for president, I pulled every string I could to help him. I believed in him. But once he got into the White House, politics got heavy. Bobby Kennedy didn't like the friends I kept—said it was a bad look. One day I was the guy at the table, the next day I was out in the cold. Hurt like hell, but that's politics. They use you until they don't.

Q: The FBI kept a thick file on you, linking you to organized crime. Did that bother you?

Sinatra: Bother me? Kid, it followed me everywhere. They had agents tailing me, digging up dirt that wasn't there. They wasted more time on me than on real criminals. I said it before, and I'll say it again—I sang for presidents, I sang for working stiffs, I sang for everybody. But somehow that made me public enemy number one. All I ever did was sing songs. If that's a crime, lock me up.

Q: You had a reputation for a fiery temper, sometimes violent. Do you think that reputation deserves?

Sinatra: Listen, I was no saint. I had a short fuse, and sometimes it got the better of me. I threw a punch or two, broke some phones, sure. But the papers loved to blow it up—made it sound like I was Al Capone. Truth is, most of the time, I was just defending myself or someone I cared about. Did I regret some of it? Yeah. But I wasn't about to let anyone push me around, not on stage, not in life.

Q: Final question. When the spotlight is gone and people remember Frank Sinatra—how do you want them to see you?

Sinatra: As a man who took the words on a page and made you believe them. That's it. Not the mob rumors, not the fights, not the politics. Just the music. Because when I sang, I wasn't hiding, I wasn't acting. I was telling the truth. And that truth will stick longer than all the rest.

Closing Statement (Narrator)

As our interview concludes, one truth becomes clear: Frank Sinatra was more than a performer; he was a force of nature. His life was marked by triumphs and troubles, but always by passion. Whether crooning a ballad in a smoky nightclub, commanding the stage in Las Vegas, or lending his voice to the silver screen, Sinatra lived unapologetically, with style and swagger. He left behind not just songs but a legacy of authenticity, resilience, and timeless artistry. Decades later, his music still resonates, reminding us why they called him simply "The Voice."

John F. Kennedy: A Timeless Conversation

Introduction: *Few figures in modern history embody both hope and tragedy as vividly as John Fitzgerald Kennedy. The 35th President of the United States, Kennedy's short time in office left an enduring mark on the nation and the world. His presidency was defined by moments of extraordinary peril and promise: the Cuban Missile Crisis, the push for civil rights, and the dream of putting a man on the moon. Known for his wit, charm, and eloquence, Kennedy inspired a generation with his call to service—urging Americans to "ask not what your country can do for you, ask what you can do for your country." In this imagined conversation, Kennedy reflects candidly on his experiences—his challenges, triumphs, and personal struggles—offering insight into the man behind the office.*

John Fitzgerald Kennedy (May 29, 1917 – November 22, 1963), also known as **JFK**, was the 35th president of the United States, serving from 1961 until his assassination in 1963. He was the first Roman Catholic and the youngest person elected president at 43 years. Kennedy served at the height of the Cold War, and the majority of his foreign policy concerned relations with the Soviet Union and Cuba.

Q: Did you ever think you would become President of the United States?

John F. Kennedy: Well, I don't think anyone grows up expecting to be President. But I was raised to believe that public service is an honorable calling, and I've always tried to do my duty wherever it led me. When I was a boy, I was just hoping to pass Latin class. The presidency wasn't exactly on my mind back then. I don't think I ever imagined it when I was young. My health was poor, and my brother Joe was the one everyone thought would carry the family's future in politics. After we lost him in the war… well, the path changed. Sometimes life chooses for you.

Q: Mr. President, looking back over your time in office so far, what would you say has been the most challenging moment of your presidency?

John F. Kennedy: Well, without question, the most serious and challenging time was the Cuban Missile Crisis in October of '62. For thirteen days, this country—and the world—stood closer to nuclear war than at any point in history. The decisions required patience, restraint, and a willingness to listen to all sides. I knew that one misstep could have meant the destruction of millions of lives. That is a burden no president ever wishes to carry.

Q: How did you manage the pressure during those days?

JFK: You rely on your team, you weigh every piece of intelligence, and you remind yourself that strength is not the same as rashness. Leadership in such moments is not about proving how tough you are; it's about proving how wise you can be. In the end, we found a peaceful solution, and I think that was the real victory.

Q: Were there other moments that you would call especially difficult?

JFK (after a pause): The Bay of Pigs operation, early in my administration, was a grave error—and the responsibility was mine. It was a bitter lesson, but it taught me never again to go forward with a decision unless I was certain of the facts and the consequences. That failure made me a stronger, wiser president when the next crisis came.

Q: How would you describe your relationship with Soviet Premier Khrushchev and with Russia in general?

JFK (leaning forward, composed): Well, as leaders of our respective countries, we have very different interests, very different systems. But it is essential that we communicate openly. The United States seeks peaceful coexistence with the Soviet Union, even as we compete ideologically and militarily.

Q: Do you trust Premier Khrushchev?

JFK (pausing, choosing words carefully): I respect him as a leader, but I do not trust blindly. The Cuban Missile Crisis, for example, reminded us of both dangers of miscalculation. Leadership in times like that requires firmness, patience, and a willingness to find common ground.

Q: Would you say the personal relationship between you and Khrushchev matters?

JFK (smiling slightly): "Personal feelings are less important than responsibility. Our duty is to protect our nations and prevent catastrophes. If that means understanding each other, negotiating carefully, and sometimes disagreeing loudly, so be it. History is watching."

Q: How was your relationship with Vice President Lindon Johnson?

JFK: My relationship with LBJ? Well, let's just say he was more comfortable wrangling votes in the Senate than turning heads in television cameras—which was fine by me. We made a good team in public, and he understood how to get things done when I needed it. While his style and mine were undeniably different, I respected what he could accomplish, and he respected the office he held—even if behind the scenes, we remained quite distinct.

Q: How was your relationship with FBI Director J. Edgar Hoover?

JFK: My relationship with Hoover… well, it was complicated. He'd been in Washington longer than just about anyone, and he knew it. He had files on senators, judges, presidents — and he let everyone know, in his quiet way, that he was not a man to be crossed.

I respected the work the FBI did on organized crime, but when it came to civil rights, Hoover dragged his feet. Bobby pushed him hard, and they were constantly at odds. Hoover didn't like being told what to do, least of all by my brother.

Did I trust him? No. Not really. I knew he kept an eye on me, maybe even on Jackie. He had too much power for one man in a democracy. But firing him would have been like kicking a hornet's nest — dangerous, unpredictable. So, we managed him instead.

To be president is sometimes to live with the people you cannot remove. Hoover was one of those men.

Q: Did you have any relations with the organized mafia?

JFK: Absolutely not. As President, my duty was to uphold the law, and my administration—particularly under Attorney General Robert Kennedy—took strong steps against organized crime in this country. The only time I ever saw the mafia was on the movie screen. And I can assure you, Mr. Brando plays the part much better than anyone I've ever met.

I never made deals with the mafia, but I can't deny that the lines blurred in politics. In 1960, there were stories about help from certain unions, certain figures in Illinois. Did I approve it? No. Did I know about every handshake made in smoky back rooms? Probably not. Bobby and I cracked down on organized crime harder than any administration before us, and that made us plenty of enemies. If there was any understanding between my campaign and those people, it didn't survive past Election Day.

Q: Mr. President, there are persistent rumors that you had numerous girlfriends both before and during your marriage. Is there any truth to that?

JFK (smiling, leaning slightly forward): Well, I suppose the press has always had a lively imagination. If I spent all my time cataloging every rumor written about me, I wouldn't have time to run the country.

Q: But surely, Mr. President, some of these stories have substance?

JFK (pausing, more serious, but calm): Look… I am a man, and I have human weaknesses. I won't pretend to be perfect. But what matters is that I've always tried to serve the American people to the best of my ability. My personal failings, whatever they may be, do not define the work we've done in this administration.

Q: And your wife, Mrs. Kennedy?

JFK (smiling warmly, almost evasively): Jackie is extraordinary. She is elegant, intelligent, and resilient. I am lucky to have her by my side, and I respect her more than words can say. Beyond that, I will leave my private life where it belongs — private.

Q: At your birthday celebration, Marilyn Monroe performed 'Happy Birthday.' Can you comment on your relationship with her?

JFK: (smiling, leaning slightly forward): Marilyn Monroe is a remarkable performer, and I was very pleased to have her sing for my birthday. She's talented, charming, and certainly made the evening memorable.

Q: Was there any personal connection beyond that performance?

JFK (with a slight pause, choosing words carefully): I maintain professional and friendly relationships with many entertainers and public figures. The rest… well, some things are best left private. That night was about celebration, not speculation.

Q: Do you have any personal fondness for her?

JFK (smiling, evasive): Marilyn is delightful, and I admire her talents. Beyond that, I prefer to keep my private life… private, as any president must.

Q: Who would you say are your closest friends?

JFK: (smiling, leaning back slightly): Well, I'm fortunate to have known some extraordinary people. My brothers, of course—Bobby and Ted—have always been my anchors. There's no one I trust more.

Q: Outside of your family?

JFK: There are a few I've known since college. Lem Billings, for one, has been a friend for decades, someone who knew me before all politics and the cameras. And Ted Sorensen, who has been not just an advisor but a confident; he's seen the best and worst of me and stuck around.

Q: Any others you'd like to mention?

JFK (smiling warmly, almost mischievously): I have a handful of others, men and women, who have been loyal and honest when I needed it most. But I've learned that true friends are rare in public life. Those who stick with you when things get difficult… those are the ones who matter.

Q: Would you say those friends have influenced your presidency?

JFK: "Absolutely. Friends provide counsel, keep you grounded, and remind you who you are beyond the office. Without them, the burdens of this job would be far heavier."

Q: With such a busy schedule, do you have any hobbies or ways to relax?

JFK (smiling): I try to make time for a few things. I love sailing, there's nothing like being out on the water with the wind in your face. It keeps you humble, especially when the boat doesn't do what you want it to.

Q: Any other activities?

JFK: "I read constantly—history, biographies, politics. I like to know what's gone by before, so I can make better decisions today. Tennis and golf are good for competition and keeping the mind sharp, and of course, spending time with my family is my favorite pastime of all."

Q So, even as President, you make time for play?

JFK (grinning): Absolutely. If a man can't play a little now and then, he'll never have the energy to do the serious work. Even in Washington, there must be time to breathe, to think, and to enjoy life a bit.

Q: Do you think your brother Robert or your son John Jr. might become President someday?

JFK: (leaning back, thoughtful): Bobby… he has the intelligence, the drive, and the courage. If the nation ever calls on him, I believe he could serve admirably. But the presidency is no small burden—it tests a man in ways most cannot imagine.

Q: And your son, John Jr.?

JFK (smiling softly, almost wistful): "John is still very young, of course. But I see in him a spark, a curiosity, and a sense of responsibility. Whether he chooses politics or some other path, I hope he grows to serve the country—or the world—in a way that matters. I also hope he avoids some of the harshness that comes with public life. It can be unforgiving."

Q: So, would you encourage them to follow in your footsteps?

JFK: I would encourage them to follow their own conscience and their own talents. Public service is a noble calling, but it is also a dangerous one. If they are willing and prepared, then perhaps one day they will take up that responsibility. If not… I hope they find a life rich in meaning and joy, wherever it leads them.

Closing Statement

As the interview draws to a close, one cannot help but sense both the weight Kennedy carried and the optimism he projected. His words reveal a leader deeply aware of the fragility of peace, the burden of responsibility, and the imperfections of human nature, his own included. Yet beneath it all was a man who believed profoundly in the power of service, the importance of courage, and the enduring promise of America.

John F. Kennedy's presidency was cut short, but his voice still echoes across time. In his reflections on leadership, friendship, family, and the uncertain future, we are reminded why his legacy endures—not only as a president, but as a symbol of possibility, vision, and hope.

Queen Elizabeth: Words Across Eternity

Narrator Introduction

She was a daughter, a wife, a mother, and a woman who never expected her destiny to unfold so soon. At just twenty-five, Elizabeth became queen, carrying a crown heavy with history and expectation. And yet, beneath the titles and ceremonies, she remained a woman of profound devotion — to her family, to her faith, and to her people. In this imagined conversation, we sit with Queen Elizabeth II not only as a monarch, but as a person who spent a lifetime balancing the weight of tradition with the quiet joys of her private life.

Elizabeth II (Elizabeth Alexandra Mary; 21 April 1926 – 8 September 2022) was Queen of the United Kingdom and other Commonwealth realms from 6 February 1952 until her death in September 2022. She had been queen regnant of 32 sovereign states during her lifetime and was the monarch of 15 realms at her death. Her reign of 70 years and 214 days is the longest of any British monarch, the second-longest of any sovereign state, and the longest of any queen regnant in history.

Q: What are your earliest memories of childhood before you became heir to the throne?

Queen Elizabeth: My earliest memories are not of palaces or of ceremonies, but of a happy childhood with my sister, Margaret, and our parents. We were quite a close family. I remember simple things: playing in the gardens, learning to ride my first pony, and the comfort of evenings spent together during very ordinary routines.

At that time, no one thought I would become heir to the throne. My uncle was king, and I was simply 'Lilibet' to my family. That sense of normality, of being a little girl with her parents and sister, is something I have always treasured. Those years gave me grounding that, later, proved invaluable when life changed so dramatically.

Q: How did your relationship with your parents and sister shape your sense of duty?

Queen Elizabeth: My parents were my greatest teachers, though not in a formal sense. My father, King George VI, showed me that duty is not about grandeur but about service and constancy, even in the most difficult times. He never expected to be king, yet when the responsibility came to him, he accepted it with humility and devotion. Watching him carry that burden left a deep impression on me.

My mother, Queen Consort, gave me a sense of resilience and optimism. She believed strongly in facing challenges with a smile, no matter how dark the circumstances.

And my sister, Margaret, brought lightness and joy into my life. She reminded me that while duty is central, one must not lose sight of the warmth and humanity that sustain us. Together, they helped me understand that service must always be balanced with compassion, and that family—both personal and nationalist at the heart of true duty.

Q: You became Queen at just 25 years old. How did you prepare yourself for such a responsibility?

Queen Elizabeth: One never truly prepares to become sovereign. Life has a way of placing responsibilities upon you before you feel ready. When my father passed away so suddenly, I was only twenty-five, and at that moment, the weight of history and duty fell upon me.

I had, of course, been observing and learning since childhood—watching my parents, understanding the rhythm of public service, and taking on small duties as a princess. But nothing can fully prepare one for the role of monarch. What gave me strength was the example of my father's steadfastness, the guidance of experienced advisers, and the unwavering support of Prince Philip.

I resolved from the very beginning that my life, whether it be long or short, would be devoted to service. That promise, made publicly at my coronation, became my compass in all the years that followed.

Q: Did you ever wish for a different life, one outside of the monarchy?

Queen Elizabeth: It is quite natural, I think, for anyone to wonder what life might have been like had circumstances been different. As a young girl, before the abdication of my uncle, I was not expected to become Queen. In those days, I imagined a quieter life — perhaps as a countrywoman, surrounded by horses and dogs, enjoying family without the weight of public duty.

But once the course of events changed, there was never a question in my mind of choosing another path. Duty is not something one selects; it is something one accepts. While my life has been bound by obligations that few can imagine, it has also been filled with opportunities to serve, to meet extraordinary people, and to witness history as it unfolded.

So although at times I wondered about that other life, I never resented the one I was given. Each life carries its own burdens and blessings, and mine has been one of service, which I have always regarded as both an obligation and a privilege.

Q: What has been the greatest challenge of balancing tradition with modernization in the monarchy?

Queen Elizabeth: The monarchy rests upon traditions that stretch back centuries, and those traditions give people a sense of continuity and stability. At the same time, the world does not stand still. Society changes, values evolve, and technology reshapes how people live and communicate.

The greatest challenge has been to maintain the dignity and symbolism of the Crown while also adapting to those changes in a way that feels natural rather than forced. For instance, the decision to televise my coronation in 1953 brought the monarchy into people's homes in a way never done before. Later, embracing travel, meeting ordinary citizens, and allowing cameras a more intimate view of the family all reflected that balance.

It has been a careful path — too much tradition, and one risks seeming out of touch; too much change, and one risks losing the very essence of what the monarchy represents. My aim has always been to uphold the institution's timeless values of service, duty, and unity, while allowing it to move gently forward with the times.

Q: You have met countless world leaders across decades. Which encounters left the deepest impression on you?

Queen Elizabeth: It has been one of the remarkable privileges of my reign to meet so many leaders from around the world — presidents, prime ministers, monarchs, and figures of history whose names are now written in the books. Each encounter carried its own meaning because each reflected the moment in time.

Some left an impression because of the gravity of the circumstances. **Winston Churchill**, for example, was my first prime minister and a guiding presence when I was new to the throne. His wisdom and deep sense of history were invaluable. **Nelson Mandela** left a very different kind of impression — one of warmth, resilience, and reconciliation after unimaginable hardship.

I also recall the quiet strength of leaders such as **Charles de Gaulle**, the optimism of **John F. Kennedy**, and the steadiness of more recent figures who worked tirelessly for peace in times of conflict.

What struck me most was not power or position, but the humanity behind it — the character, humility, and sense of purpose that great leadership requires. Those are the qualities that endure long after the titles fade.

Q: How did you personally navigate the political turbulence of the Cold War, decolonization, and global conflicts?

Queen Elizabeth: As sovereign, my role has never been to make political decisions but to provide continuity, stability, and a symbol of unity above the changing tides of government. During the Cold War, when the world lived under the shadow of nuclear conflict, I believed it was important that the monarchy stand as a steady presence — calm, impartial, and reassuring to the people.

Decolonization brought with it profound changes. Countries that had once been part of the British Empire chose independence, and it was my task to help nurture a different kind of relationship through the Commonwealth. I viewed this not as a loss, but as an opportunity to forge new bonds of friendship and cooperation built on equality and shared values.

In times of war or global crisis, my duty has been to offer encouragement, to listen, and to remind people of our shared humanity. While leaders debated and armies clashed, the monarchy's task was to uphold dignity, to extend the hand of friendship across divides, and to embody the hope that peace is always possible. That constancy, I felt, was the most valuable contribution I could make.

Q: You worked with many British Prime Ministers. Who were your favorite PMs?

Queen Elizabeth: It would not be quite proper for me to single out 'favorites,' for each Prime Minister brought their own qualities and faced unique challenges of their time. My role was not to prefer one over another, but to listen, to advise when sought, and to support them in their duty to govern.

That said, I will admit that my very first Prime Minister, **Winston Churchill**, held a special place in my heart. I was new to the throne, and he, with his immense experience and gravitas, was a source of reassurance. His stories, his sense of history, and his deep devotion to the country left a lasting impression.

In later years, I greatly valued the candor and warmth of **Harold Wilson**, with whom I had a particularly good rapport. And I admired **Margaret Thatcher**'s formidable sense of determination, even when our views differed.

What I found most striking across them all was not ideology, but dedication. From **Clement Attlee to Tony Blair** and beyond, they shouldered enormous burdens, and I respected their commitment to serving the nation. My task was always to be a constant — a sounding board, a confidante, and a reminder that institutions endure even as governments change.

Q: Did you have close personal friends outside the royal family?

Queen Elizabeth: Yes, though my position did not always allow for friendships in the ordinary sense. A monarch's life is lived so publicly, and one must always be mindful of discretion and duty. But within those boundaries, I did form bonds of trust and affection with people beyond my family.

Some were lifelong companions from childhood; others were members of the household who became confidantes over the years. I also valued friendships formed with those who shared my interests — horses, dogs, the countryside. In such a company, there was a sense of ease, of being able to speak and laugh without ceremony.

These friendships, though often private and away from the gaze of the world, were important to me. They reminded me that behind the crown, there is still a person in need of companionship and warmth, like anyone else. They gave me perspective, balance, and at times, solace

---

Q: What role did Prince Philip play in supporting you throughout your reign?

Queen Elizabeth: Prince Philip was my strength and stay for more than seventy years. From the moment we married, he was not only my husband but also my partner in every sense of the word. He gave me unwavering support, often quietly and without recognition, so that I could fulfill the duties of my position.

He had a remarkable ability to see matters clearly and directly. His sense of humor, his energy, and his curiosity about the world helped to keep me grounded. He did not seek the spotlight, yet he carried out countless engagements with dedication, and his interest in causes such as conservation and technology left their mark on the country.

Most of all, he understood the unique weight of the role I carried, and he bore the demands of royal life with resilience. His presence gave me reassurance, and his companionship gave me joy. It is difficult to imagine my life — and indeed my reign — without him by my side.

Q: How was your relationship with Princess Diana? Public perception was that it was very controversial.

Queen Elizabeth: Diana was a remarkable young woman who came into the royal family at a very young age and under great public attention. She had a natural warmth and a gift for connecting with people, qualities that endeared her to the nation and to the world.

Our relationship, like many within families, was not without its complexities. The pressures of public life, the intense scrutiny of the press, and the strains upon her marriage all contributed to challenges that were difficult for everyone involved. At times, misunderstandings arose, as they do in any family.

Yet I always recognized Diana's devotion to her sons and her unique ability to reach those who felt overlooked or unheard. Her charitable work and her compassion left a legacy. While people may remember controversies, I prefer to remember the good she brought to the world and the love she gave to her children, who carry forward her spirit

Q: How do you reflect on the monarchy's relationship with the press, especially during challenging times?

Queen Elizabeth: The press has always played a vital role in informing the public, and with the monarchy being a public institution, scrutiny has long been part of our existence. In many ways, the press has helped bring the work of the Crown closer to people — showing our engagements, our charitable efforts, and the traditions that bind us.

But there have also been times when the intensity of attention became difficult, particularly for members of my family. The line between public duty and private life is not always easy to maintain, and the press can sometimes blur that distinction.

I have always believed that one must not be unduly swayed by headlines, whether critical or complementary. Institutions endure because they are steady. My task has been to remain consistent, to keep faith with my oath of service, and to trust that, over time, people see the value of continuity.

The monarchy's relationship with the press has been, at times, challenging, but it has also been essential in allowing the Crown to remain visible, relevant, and accountable to the people it serves.

Q: Do you believe the monarchy will continue to play a vital role in the future?

Queen Elizabeth: The institution of monarchy has endured for centuries because it adapts while holding fast to its essential purpose: service, stability, and continuity. The world changes rapidly — governments come and go, technology reshapes daily life, and societies face new challenges — but people still find reassurance in a symbol that stands above politics and personal ambition.

I have always believed that the monarchy's role is not to dictate, but to represent — to embody the unity of the nation and the Commonwealth, to give people a sense of belonging and identity. As long as it continues to serve with humility and dedication, it will remain meaningful.

Of course, the monarchy must always evolve. Each generation will bring its own character and approach, as it should. But the values of duty, constancy, and service do not go out of fashion. If those remain at the heart of the Crown, I believe it will continue to play a vital role well into the future.

Q: What advice would you give to future generations?

Queen Elizabeth: I would say to future generations: never lose sight of the values that endure — kindness, integrity, and a sense of duty to one another. The world will change in ways none of us can fully imagine, but human character will always matter more than circumstance.

Remember that service to others is not a weakness, but a strength. It is through giving — of time, of compassion, of understanding — that communities are built, and peace is sustained.

Do not be discouraged by the difficulties of your age. Each generation faces its trials. What matters is the spirit with which you meet them: with courage, with faith, and with the knowledge that even the smallest acts of goodness can ripple far beyond what you see.

If you hold fast to those principles, the future, whatever form it takes, will be one of hope.

Narrator Closing

As our conversation draws to a close, we are reminded that Queen Elizabeth's story is not simply about crowns and kingdoms, but about endurance, sacrifice, and love of duty. She gave her entire life to service, yet also cherished the roles of mother, grandmother, and great-grandmother. Though the world has said farewell to its longest-reigning queen, her presence lingers — in the institutions she preserved, in the family she nurtured, and in the memories of millions who found comfort in her constancy.

Indira Gandhi: Strength and Sacrifice

Narrator Introduction

Indira Gandhi (1917–1984), daughter of India's first Prime Minister Jawaharlal Nehru, remains one of the most complex and influential figures in modern history. Serving as Prime Minister for nearly 16 years, she was known for her boldness, her centralizing leadership, and her determination to shape India's destiny. Loved and reviled in equal measure, she left a legacy that still sparks debate today. In this imagined conversation, Indira Gandhi reflects power, leadership, and her place in history.

Q: You grew up as the daughter of Jawaharlal Nehru, India's first Prime Minister. How did that shape your path to leadership?

Indira Gandhi: Politics was the air I breathed from childhood. My father taught me that public service is both an honor and a burden. I did not seek leadership to mimic him, but because I felt a deep duty to continue building the India he and Mahatma Gandhi envisioned—independent, united, and strong.

Q: You were often described as the "Iron Lady of India." Do you agree with that description?

Indira Gandhi: If by "Iron Lady" they mean that I did not bend easily to pressure, then yes. Leadership requires resolve. India is vast, diverse, and often chaotic; without firmness, governance would descend into paralysis. Yet I was also deeply sensitive, perhaps too much so at times. Strength is not an absence of feeling, but the ability to act despite it.

Q: The Emergency (1975–77) remains one of the most controversial periods of your rule. How do you justify that decision?

Indira Gandhi: The Emergency was not declared out of vanity or whim. At that time, India faced turmoil—economic instability, protests, strikes, and threats to internal security. I believed democracy itself was at risk of unraveling. Was it a harsh step? Yes. Do I regret the excesses that occurred under its shadow? Certainly, but I maintain that my intention was to preserve India's unity, not to destroy its democracy.

Q: You were the only woman to lead India as Prime Minister. Did your gender make leadership more difficult?

Indira Gandhi: In India, people respected me not because I was a woman, but because I was a leader. That said, being a woman in politics meant constant underestimation. Many assumed I would be a puppet when I first took office. I proved them wrong. I never defined myself by my gender, though I understood the symbolic importance of a woman leading the world's largest democracy.

Q: Operation Blue Star in 1984 is another defining and tragic moment of your tenure. How do you reflect on it now?

Indira Gandhi: It was one of the most painful decisions of my life. The Golden Temple had been turned into a fortress by extremists, and law and order were collapsing in Punjab. As Prime Minister, I had to act to protect the nation. But I knew the cost would be immense. It cost me dearly—politically, morally, and personally. And ultimately, it cost me my life.

Q: Looking back, what do you consider your greatest achievement?

Indira Gandhi: The Green Revolution that secured India's food supply, the strengthening of our defense, and the assertion of India's independence on the global stage—these gave me pride. But perhaps my greatest achievement was proving that India, though poor and diverse, could act with dignity and decisiveness among the great nations of the world.

Q: Mrs. Gandhi, you met countless world leaders during your time in office. Which of them impressed you the most?

Indira Gandhi: Each leader I encountered reflected their nation's character and their own temperament, and I learned something from nearly all of them. Yet some stand out more clearly in my memory.

From the East, I admired leaders like Zhou Enlai of China. He had a sharp intellect and a subtlety that was both fascinating and formidable. One could not underestimate him. From the West, I respected Charles de Gaulle — his sense of history, his independence of spirit. He understood what it meant to protect the dignity of a nation, even when it was difficult.

I also valued my meetings with leaders of the developing world, who, like me, carried the burden of building nations still finding their footing after colonial rule. Julius Nyerere of Tanzania, for example, impressed me with his sincerity and his vision of self-reliance. These were not leaders of great military powers, but of great moral conviction.

Among the superpowers, I dealt with both American and Soviet leaders. I found some Americans impatient, always in search of quick results, while the Soviets were often rigid. But in both, I recognized the weight of their responsibilities. With the Soviets, particularly Brezhnev, I shared a mutual understanding of strategic necessity, though we came from different worlds.

Ultimately, the leaders who impressed me most were not always the most powerful, but those who carried themselves with dignity, integrity, and an awareness of history. For me, respect is not given to strength alone, but to wisdom.

Q: Mrs. Gandhi, you devoted so much of your life to politics and public service. Could you tell us about your family life? Did you have a love of your life?

Indira Gandhi: My public life was very visible; my private life, I tried to keep quiet. I married Feroze Gandhi when I was young. He was not only my husband but my friend and partner in ideals. We shared a belief in an India that could be fairer, more just, and more compassionate. His death in 1960 left a void in me that never truly healed. If I had a 'love of my life,' it was Feroze — though our marriage, like any, had its tensions, he was the one person who saw me not as a leader or a symbol, but as Indira.

After his passing, my life became almost entirely about my children and my country. My sons, Rajiv and Sanjay, were my anchors and my pride. I poured myself into them, and into India, sometimes at the cost of my own peace. In a sense, my 'family life' became indistinguishable from my work, because I felt a duty not just to my family but to 700 million people. But my most intimate happiness was always with my children — and my greatest heartbreaks were there too.

Q: Mrs. Gandhi, you spent so much of your life in politics. Do you have any close friends?

Indira Gandhi: True friendship is rare in politics. Power attracts many people, but sincerity far fewer. Throughout my life, most of those around me were colleagues, allies, or critics, not intimate friends. Yet I was not without companionship.

As a young woman, I had friends from my time studying in Switzerland and at Oxford — people who knew me simply as Indira Nehru, not as a prime minister's daughter or a future leader. Later, I valued the quiet loyalty of a few lifelong companions, like Pupul Jayakar, who was both an adviser and a confidante. She understood me as a woman, not just as a politician.

Still, leadership is isolating. When you are responsible for a nation, you guard your heart. You learn to be courteous, to be open, but also to protect yourself. My closest and deepest bonds were with my family — my father, my sons. Beyond them, only a handful of people truly knew me.

Q: And what do you hope future generations will remember about you?

Indira Gandhi: That I acted with courage, even when my choices were questioned. That I placed India above myself. And that in moments of crisis, I did not flinch. History may judge me harshly, but I hope it will also remember my devotion to my country.

Narrator Closing

Indira Gandhi's life was one of contradictions—nurturing yet ruthless, democratic yet authoritarian, visionary yet controversial. She shaped India with her strength and will, and in doing so, she became a symbol of both progress and peril. Her voice reminds us that leadership is never simple, and that history's verdict is always complex.

Pope Francis: A Meeting of Minds

Narrator's Introduction

In the heart of Vatican City resides a spiritual leader whose influence extends far beyond the walls of St. Peter's Basilica. Born Jorge Mario Bergoglio in Buenos Aires, Argentina, he became the first Jesuit pope, the first from the Americas, and the first to choose the name Francis, honoring the saint known for humility and love of the poor. His papacy has been marked by compassion, openness, and a call for justice on issues ranging from poverty and migration to climate change and human dignity. Today, we embark on an imagined conversation with Pope Francis — a man whose voice has resonated with millions across faiths, cultures, and nations.

Pope Francis (born **Jorge Mario Bergoglio**; 17 December 1936 – 21 April 2025) was head of the Catholic Church and sovereign of the Vatican City State from 13 March 2013 until his death in 2025. He was the first Jesuit pope, the first Latin American, and the first born or raised outside Europe since the 8th-century Syrian pope Gregory III.

Q: Did you ever think that one day you would become Pope?

Pope Francis: No, no, never! I never thought I would become Pope. I was happy just being a parish priest, being close to the people. When it happened, it was a real surprise — I would say even a shock. God likes to surprise us, you know. I simply tried to accept it with trust and a little trembling… and maybe still today I ask myself, *'Why me?*

Q: What was the main reason you became a priest?

Pope Francis: Why did I become a priest? Ah… because God caught me by surprise! (laughs) I had other plans — I was studying, I had friends, I thought about starting a family. But God had a better plan. He tapped me on the shoulder in the confessional, and from that moment, I couldn't escape. Priesthood is not a career or a choice we make like any other. It is a response to a call, to a love that is greater than us. I became a priest to be a servant of that love.

Q: How would you define your style as Pope compared to prior popes?

Pope Francis: Ah, style? I don't know if I have a style! (laughs) Maybe it's a bit more informal... I like to speak simply, greet people, pick up the phone, and call someone who wrote me a letter. Some people say, *'This is unusual for a pope,'* but for me, it's normal. The Pope is called to serve, not to be served. My way is to put mercy at the center and to remind the Church that it must go out, not stay closed in on itself.

Q: What do you consider your main legacy?

Pope Francis: Legacy is not for me to decide. I leave that to God and to history. But if people remember me, I hope it is for mercy — for insisting that God never tires of forgiving, and that the Church must go out to the streets, to the poor, to those who are forgotten. If that remains, I will be grateful.

Q: You often speak about the poor. Why is this so central for you?

Pope Francis: I always say the poor are at the center of the Gospel. If we forget them, we forget Jesus. They are not numbers or statistics — they have faces, names, stories. To serve the poor is not charity in the sense of 'giving from above', but justice, because they have the same dignity as all of us.

Q: What about the environment? Why do you place such importance on caring for creation?

Pope Francis: The earth is our common home. If we don't protect it, who will? It is not only about nature; it is about people — especially the poor, who suffer first when the environment is damaged. Small actions matter, such as reducing, recycle, and respecting the world around you.

Caring for the environment is caring for our brothers and sisters, and for future generations.

Q: What message do you have for young people today?

Pope Francis: To the young, I say: do not be afraid. Make noise, go out, don't live your life on the couch. Take risks for something great! The world needs your energy, your dreams, your courage. The Church needs you.

Q: Can you tell us a little about your daily life?

Pope Francis: My life is quite simple. I live in the Casa Santa Marta, not in the Apostolic Palace, because I prefer being close to others. I celebrate Mass, I pray, I work, I meet people. Sometimes I even make my own coffee! I try to live as simply as I can, because simplicity keeps the heart free.

Q: As a priest, and later as a cardinal and Pope, did you have some close friends?

Pope Francis: Yes, I have been blessed with close friends throughout my life. As a priest, I had companions with whom I could pray, talk, and share my joys and struggles. Even now, as Pope, I have friends — fellow priests, bishops, and people I have known for many years. Friendship is a gift from God, and it keeps the heart humble and joyful. A friend can make you laugh, remind you of what matters, and sometimes even scold you.

Q: Did you have enemies?

Pope Francis: I try not to think in terms of enemies. Of course, there are people who disagree with me or with my decisions. But I pray for them, and I try to understand them. Jesus tells us to love our enemies and pray for those who persecute us. The Church is called to reconciliation, not division. Those who oppose me or the Church are still children of God, and I entrust them to His care.

Q: What is your position on sexual harassment in Catholic churches?

Pope Francis: These scandals are a deep shame for the Church and for all of us. There is no place for abuse. Every accusation must be taken seriously, investigated carefully, and justice must be done. Abuse is a sin, a crime, and a betrayal of trust. We cannot hide it or cover it up. The Church must act decisively, remove those who abuse, and walk with the victims so they know they are not alone. Abuse wounds the Body of Christ. We must pray, act justly, and never forget that God's mercy is for the repentant, but justice is for the victims.

shutterstock.com · 2575693941

Q: What is your position on LGBTQ people, including children?

Pope Francis: Every person is a child of God and deserves to be treated with dignity, respect, and love. The Church must welcome everyone, without exception, and accompany them with care. No one should ever feel rejected by God or by the Church. I often say: If a person is gay, who am I to judge? Our mission is to walk with people, to support them and their families, and to make sure they know they are loved. Children, like all children, must be protected and cared for, always. God's mercy is for all. We are called to love, not to condemn.

Q: You met lots of world leaders. Could you tell us your opinion about some of them?

Pope Francis: Meeting world leaders is always a responsibility and an opportunity. I do not comment on politics, but I can say that some leave a deep impression through their commitment to justice, peace, and the common good. Others may struggle, and I pray for them, because governing is a great responsibility and often a heavy burden. I try to see the person beneath the office, to understand their heart, and to encourage them to care for the poor, the vulnerable, and our common home.

Q: Please be more specific. Which world leaders left a deep impression on you?

Pope Francis: Rather than naming many individuals, I reflect on those who acted with courage, mercy, and justice. Pope John Paul II inspired me as a global leader; I have met presidents and prime ministers who follow that path, caring for their people and speaking for peace. Those encounters stay in my heart. I remember leaders such as Angela Merkel, who carried her responsibilities with patience and prudence; Justin Welby, though a religious leader, showed me the power of dialogue; and the presidents of small nations who devote themselves to protecting the vulnerable in difficult circumstances.

Q: Do you believe that one day women may become cardinals, and that one day we may have a woman pope?

Pope Francis: Women already play an essential role in the Church, and their contribution is immense. They bring wisdom, leadership, and care in every area. As for becoming cardinals, it is possible that in the future the Church may recognize women in new ways of governance. But the papacy, as we know it today, is reserved for men. That does not diminish the vocation, influence, or dignity of women in any way. The Spirit continues to guide the Church toward justice, recognition, and inclusion. Women are already indispensable in the Church's mission, and their role must grow in wisdom and authority.

Narrator Closing

As our dialogue with Pope Francis draws to a close, we are left with reflections not only on faith but on humanity itself. His words remind us that leadership is not about power, but about service; not about privilege, but about responsibility to one another and to the Earth we share. Whether one views him as a religious leader, a moral guide, or simply a fellow human striving for a kinder world, Pope Francis continues to challenge us to build bridges where walls divide. In the end, his message is clear: hope, compassion, and mercy are not lofty ideals, but the very essence of a well-lived life.

Napoleon Bonaparte: An Imagined Conversation

Napoleon Bonaparte remains one of history's most fascinating and polarizing figures. Rising from obscure Corsican roots to become Emperor of the French, he reshaped Europe with his military genius, political ambition, and sweeping reforms. Admired for his brilliance yet condemned for his wars, he was both a conqueror and a visionary. In this imagined interview, we revisit his story through his own words.

Napoleon Bonaparte (born **Napoleone di Buonaparte**; 15 August 1769 – 5 May 1821), later known by his regnal name **Napoleon I**, was a French general and statesman who rose to prominence during the French Revolution and led a series of military campaigns across Europe during the French Revolutionary and Napoleonic Wars from 1796 to 1815. He led the French Republic as First Consul from 1799 to 1804, then ruled the French Empire as Emperor of the French from 1804 to 1814, and briefly again in 1815. He was King of Italy from 1805 to 1814 and Protector of the Confederation of the Rhine from 1806 to 1813.

Q: You were born on Corsica, a small island. Did you ever imagine that one day you would rule much of Europe?

Napoleon: As a boy, I dreamed of greatness, but who could have foreseen such a destiny? Corsica was small, poor, and overlooked. Yet I believed that through discipline, intellect, and will, a man could rise above his origins. France gave me that chance—and I seized it with both hands.

Q: Your rise to power was meteoric. What do you believe sets you apart from other generals and politicians of your time?

Napoleon: Clarity. In the chaos of battle, or the confusion of politics, I saw what others did not. I believed in decisive action, in bold strokes. Where others hesitated, I moved. Fortune favors the daring, and I was never afraid to stake everything on a single throw of the dice.

Q: You crowned yourself Emperor in 1804. Why did you not allow the Pope to crown you, as tradition dictated?

Napoleon: Because my crown was not given by the Church, nor by tradition—it was seized by my own hand. By crowning myself, I declared that my power came from my achievements, not from divine sanction. I was the maker of my own destiny.

Q: In your life, you led so many battles, across Europe and beyond. Why was war so constant in your destiny?

Napoleon: Because Europe left me no choice. I was born in revolution, forged in fire. France was surrounded by kings who feared our ideals of liberty and equality. They sought to strangle the Republic in its cradle. To defend France, I had to fight. And once I fought, I learned that victory itself brings new enemies. Each triumph demanded another, until war became both my weapon and my burden.

But do not think I loved war for its own sake. I loved victory — the clarity of command, the discipline of men united, the thrill of seeing a plan unfold as I had foreseen it. On the battlefield, all masks fall away; one confronts truth in its rawest form. That is what drew me, repeatedly.

Yet I also fought because war was the language of power in my age. To secure my throne, to spread the laws and reforms I believed in, I had to wield the sword. My enemies would not yield words. If Europe had allowed me peace, I would have given them prosperity. But as they gave me war, I gave them empire.

Still, I know the cost. My victories built a legend, but they also consumed a generation. If I am remembered as a man of battles, it is because my life and my age were inseparable from war.

Q: You are remembered as both a military genius and a tyrant. How do you see yourself?

Napoleon: History will argue forever, I suspect. To some, I was a tyrant, to others, a liberator. I introduced laws that endure—the Napoleonic Code, which gave structure to modern justice. I shattered feudal privileges. Yet I also waged wars that consumed lives. I was a man of contradictions—brilliant and flawed, visionary and ruthless.

Q: Many say your greatest mistake was invading Russia. Do you agree?

Napoleon: Yes. Russia was my graveyard. Not in the battles, but in the winter, the hunger, the endless retreat. My army was not defeated by Russian steel, but by cold, starvation, and distance. I underestimated the vastness of that land and the resolve of its people. That was my greatest folly.

Q: History records that you had Pope Pius VII arrested and held prisoner. How do you explain such an act against the head of the Catholic Church?

Napoleon: The Pope may be the vicar of Christ, but I was the Emperor of the French. France had bled through revolution, and I would not allow any power — spiritual or temporal — to divide the nation or challenge my authority. Religion has its place, but so does sovereignty. When Pius VII defied me, when he sought to undermine my decrees and alliances, I acted as any ruler must: I secured my throne.

It was not cruelty, but necessity. I had no wish to destroy the Church, only to bend it to the needs of France and of the empire. Did I not bring the papacy back to Paris to crown me? Did I not reconcile France with Catholicism through the Concordat? But the Pope forgot that while his kingdom was of heaven, mine was of earth — and in this world, it was I who commanded.

History may judge me harshly for laying hands on the Holy Father, but remember: my duty was not to Rome, but to France. An emperor cannot serve two masters. I chose my people over the papacy, my authority over compromise. That is why I acted as I did.

Q: What is your opinion of women? Did you have a love of your life?

Napoleon: Women, like men, are forces of nature. They inspire, they wound, they command, they console. In my youth, I often spoke like a soldier — bluntly, foolishly — and I am aware that some of my words about women have been remembered unkindly. But in truth, my life was shaped by them: my mother's iron will, my sisters' ambitions, the elegance of the salons of Paris. I admired intelligence and spirit as much as beauty.

As for the love of my life — yes. Josephine. She was not perfect, nor was I, but together we were fire and smoke. She captivated me in a way no one else ever did. Even after I divorced her for reasons of state, she remained in my heart. When I learned of her death at Malmaison, even on campaign, I wept like a man undone. Others followed, of course — Marie-Louise gave me an heir, and there were liaisons before and after — but Josephine was my true passion. She was my first Empress, and in my heart, she was my last.

Q: Did you have some close friends during your rise and rule?

Napoleon: Friendship, like loyalty, is rare among those who hold power. Many who called themselves my friends were, in truth, my allies—or my rivals waiting for their turn. Yet there were a few men I trusted, and in my world, that was no small thing.

In my youth and early campaigns, I had a circle of devoted companions: Jean Lannes, who rose from humble origins to become one of my finest marshals—he was more than a general, he was a brother in arms. Duroc, my Grand Marshal of the Palace, was a man of unshakable honesty; his death in 1813 was a wound I felt deeply. I also trusted Caulaincourt, who negotiated for me even in the darkest hours, and Berthier, who understood my mind on the battlefield almost before I spoke.

But as Emperor, true friendship became almost impossible. Power isolates. Every kindness is weighed against ambition. I learned to be my own counsel, and perhaps that is why my last years were so lonely.

Napoleon's Inner Circle

Though he commanded millions and ruled as Emperor, Napoleon Bonaparte often said that true friendship was a luxury denied to men of power. Yet behind the grandeur of court and the thunder of battle, a handful of loyal companions stood by his side — men who earned his trust not through flattery, but through devotion, honesty, and shared struggle. Here are the figures who formed Napoleon's closest circle.

Marshal Jean Lannes (1769–1809) was a fellow soldier of humble birth who rose alongside Napoleon to become one of his most brilliant marshals. Fiercely brave, blunt, and loyal, he was one of the few men who could speak to Napoleon without fear. Napoleon once said of Lannes: *"I* *found him a pygmy, and I lost him a giant."* His death at Aspern-Essling devastated Napoleon, who grieved not only for a marshal but for a true friend.

General Géraud Duroc (1772–1813) As Grand Marshal of the Palace, Duroc managed Napoleon's household, but he was more than a courtier — he was a confidant. Calm, loyal, and incorruptible, Duroc earned the emperor's respect in both personal and official matters. When Duroc was killed in battle in 1813, Napoleon wept openly, saying, *"Everything can be replaced, except for Duroc."*

Armand de Caulaincourt (1773–1827) Caulaincourt served as Napoleon's trusted ambassador to Russia and later as Grand Equerry. Unlike many courtiers, he dared to tell hard truths. Caulaincourt urged Napoleon not to invade Russia, warning of disaster. Though ignored, his honesty marked him as a rare true friend. During Napoleon's downfall, he remained at his side, negotiating honorably for his master's fate.

Marshal Louis-Alexandre Berthier (1753–1815) was not the most charismatic of the marshals, but his organizational brilliance made Napoleon's campaigns possible. He understood the emperor's methods so deeply that he could draft orders exactly as Napoleon intended. Berthier's loyalty and efficiency freed Napoleon to focus on strategy. The emperor once admitted: *"Without Berthier, I would have been lost."*

Napoleon remained a man often surrounded by opportunists and flatterers. Yet in **Lannes, Duroc, Caulaincourt, and Berthier**, he found what he valued most: courage, loyalty, honesty, and competence. These men were not merely servants of an empire; they were companions who stood beside a man burdened by destiny.

Q: You were exiled twice, first to Elba, then to Saint Helena. How did you endure the final years of your life?

Napoleon: A caged eagle still dreams of the sky. On Elba, I plotted a return—and I did return, if only for a hundred days. But on Saint Helena, there was no return. I lived with my memories, my regrets, and my pride. I wrote, I dictated, I tried to shape how posterity would see me. Perhaps that was my final campaign.

Q: Looking back, what do you believe was your true legacy?

Napoleon: It is not the battles that fade in the dust of history. My true legacy lies in the laws I left, the meritocracy I defended, the belief that a man may rise not by birth but by talent. If Europe remembers me only as a conqueror, then it forgets half of who I was.

Closing Statement:

Napoleon Bonaparte's shadow still looms over Europe and the world. To some, he was a despot who brought misery to millions. To others, a visionary who modernized societies showed that greatness could come from anywhere. Like the empire he built, his life was brilliant, turbulent, and short-lived—but unforgettable.

Princess Diana: Conversation Beyond Reality

She was more than a princess — she was a young woman who captured the hearts of millions with her warmth, vulnerability, and courage. Diana Spencer stepped into the royal spotlight at just nineteen, yet she remained deeply human, revealing both her struggles and her strength to a world that could never look away. She gave her love most freely as a mother and as a champion for the forgotten ones: the sick, the poor, the voiceless. In this imagined conversation, we meet Diana not as a distant figure of royalty, but as a woman who longed for authenticity, compassion, and a life of her own.

Diana, Princess of Wales (born **Diana Frances Spencer**; 1 July 1961 – 31 August 1997), was a member of the British royal family. She was the first wife of Charles III (then Prince of Wales) and the mother of Princes William and Harry. Her activism and glamour, which made her an international icon, earned her enduring popularity.

Q: What was the most difficult part of transitioning from private life into the royal family?

Princess Diana: I think the hardest part was losing the sense of being ordinary. Overnight, my life became public property. Every smile, every word, every outfit was scrutinized. I wanted so much to do good with the platform, but at the same time, I often felt very lonely, because there wasn't much guidance on how to navigate it. It was a huge adjustment — not just for me, but for my sense of self.

Q: Was there anyone in the royal family who helped you through that transition?

Princess Diana: In the beginning, there was a great deal of kindness shown to me, but I wouldn't say there was much guidance. It's a world with its own rules and traditions, and I was expected to simply know them. I think I felt quite isolated at times. The people who really helped me through were often outside the family — my close friends, and later, my boys. They gave me strength when I needed it most.

Q: Did you ever feel truly accepted by the royal family?

Princess Diana: I think 'accepted' is a complicated word. I was welcomed formally, of course, but I often felt more like a figure than a member of the family. There were times when I felt embraced, and there were times when I felt very much on the outside. What I longed for was warmth and understanding, and that didn't always come easily in that environment. So, no, not always. I often felt I had to create my own space, my own way of belonging.

Q: Many have speculated about difficulties in your marriage. What do you believe went wrong between you and Prince Charles?

Princess Diana: I think what went wrong was that we were two very different people with very different needs. I wanted love, warmth, and partnership, while Charles had his own path, his own interests, and pressures. From the beginning, there were three of us in that marriage, and that made it very difficult. I don't believe either of us set out to hurt the other, but the lack of understanding and constant public scrutiny made things almost impossible to repair.

Q: How do you respond to the constant rumors about your private life?

Princess Diana: The rumors were painful, of course, especially when they weren't true. It's very hard to defend yourself when the story has already been written for you. But I learned, over time, that the only thing I could do was live as honestly as I could. People will always speculate, but I knew my boys would one day be old enough to see who I really was. That thought gave me peace — because I wanted them to know the truth about their mother, not the headlines.

Q: Did the media ever get something right about you, or did it always feel distorted?

Princess Diane: I think sometimes they caught glimpses of the truth — the part of me that genuinely loved meeting people, holding someone's hand in a hospital, or kneeling to talk to a child. That was real, and I was glad when the media showed that side. But too often, it was distorted, because the need for a headline outweighed the need for truth. They liked to create a character — the fairytale princess, or the troubled woman — but I was neither of those alone. I was simply me, a mixture of both light and struggle.

Q: You did have romantic relationships with other men?

Princess Diana: Yes, I did have relationships, but I think what people often forget is that I was searching for love and companionship, just like anyone else. After feeling very lonely for a long time, I wanted to be with someone who saw me as a woman, not as a role or an image. Those relationships were very private to me, even if the world tried to make them into scandals.

Q: Were you ever able to find the kind of love you had always hoped for?

Princess Diana: I don't know if I ever found the lasting love I dreamed of. I certainly found moments of it — times when I felt cherished and understood. But I suppose what I longed for most was a love that felt safe, steady, and unconditional. In many ways, the truest love I ever knew was from my boys. They gave me the kind of pure, unquestioning affection I had always hoped for, and that sustained me through so much.

Q: Do you think the royal family is out of touch with ordinary people?

Princess Diana: I think the royal family has a great sense of duty, but yes, at times it can seem out of touch. The formality, the traditions — they create a distance. What I always tried to do was bridge that gap, whether by sitting on the floor with a child in a hospital or holding the hand of someone society had forgotten. People don't just want to see crowns and ceremonies; they want to feel that they matter, that they are seen. That's where I felt my role was — to bring warmth and humanity into the picture.

Q: Was there ever a moment when you felt like leaving the royal life entirely?

Princess Diana: Yes, there were times when I felt overwhelmed enough to want to walk away completely. The pressure, the scrutiny, the loneliness — it could feel unbearable. But I also knew that leaving wasn't simple, because my boys were at the heart of it all. They needed stability, and I wanted to protect them. In a way, I did leave parts of the royal life behind by choosing to live more independently later. I carved out a space where I could be Diana, not just the Princess of Wales.

Q: How would you describe your relationship with your sons, William and Harry, and what values do you hope they carry forward?

Princess Diana: My boys are everything to me — they are my anchor, my joy, and my greatest achievement. From the beginning, I wanted them to grow up knowing love, warmth, and compassion, not just protocol and duty. I've always tried to show them the real world; to take them beyond palace walls so they could see how other people live, and understand the struggles as well as the privileges.

The values I hope they carry forward are kindness, empathy, and the courage to be themselves. Titles come and go, but character lasts forever. If they can treat people with dignity and live with honesty, then I will feel I've done my job as their mother.

Q: Have there been people you trusted deeply who later betrayed that trust?

Princess Diana: Yes, sadly, there were people I trusted who later betrayed that trust. It was one of the hardest lessons I had to learn about, that not everyone around me cared for Diana the person; some were more interested in Diana the title. It hurt deeply because I've always worn my heart on my sleeve, and I believed in loyalty. But over time, I came to understand that betrayal says more about the betrayer than it does about me. It made me value true friendship more, and I held on tightly to the few who stood by me, no matter what.

Q: What has been the most misunderstood aspect of your life?

Princess Diana: I think the most misunderstood aspect of my life is who I truly am as a person. The media often portrayed me as fragile, troubled, or merely a figure in a fairy tale. In reality, I've faced challenges, yes, but I've also tried to live with courage, compassion, and a desire to help others.

People see the headlines, the photographs, the rumors, but they rarely see the heart behind them. I hope that one day, people will understand that I am more than the story written about me — I am someone who cares deeply and tries, in my own way, to make a difference.

Q: If you could change one thing about your journey so far, what would it be?

Princess Diana: If I could change one thing, I think I would have learned earlier how to protect my own heart while still giving love so freely. I often gave too much of myself, and sometimes that left me vulnerable. But I wouldn't change the love I gave, the friendships I cherished, or the work I devoted myself to. Every experience, even the painful ones, taught me something about courage, compassion, and the importance of being true to oneself. Life is never perfect, but it is always meaningful if you try to make a difference.

Narrator Closing

As our dialogue with Princess Diana comes to an end, we are left not with images of crowns or ceremonies, but with the memory of a woman who lived with a rare sincerity. Her laughter, her kindness, and even her pain continue to touch the world she left behind. Though her life was cut short, her legacy endures — in the causes she championed, in the sons she loved so fiercely, and in the millions who still carry her spirit in their hearts. Diana's story reminds us that true beauty lies in empathy and that even the most fragile souls can leave the most powerful mark on humanity.

Muhammad Ali: An Interview with the Greatest

Narrator's Introduction:

He called himself The Greatest — and proved it to the world. Muhammad Ali was more than a boxing champion. He was a poet, a showman, a revolutionary, and a humanitarian. From Louisville's streets to the world stage, Ali shook up the world not just with his fists, but with his words and his principles. What follows is a conversation that captures his spirit — his fire in the ring, his courage in the face of opposition, and his enduring wisdom.

Muhammad Ali, born Cassius Marcellus Clay Jr. (January 17, 1942 – June 3, 2016), was an American professional boxer and social activist. A global cultural icon widely known as "**the Greatest**", he is often regarded as the greatest heavyweight boxer of all time. He held the Ring magazine heavyweight title from 1964 to 1970, was the undisputed champion from 1974 to 1978, and was the WBA and Ring heavyweight champion from 1978 to 1979. In 1999, he was named Sportsman of the Century by Sports Illustrated and the Sports Personality of the Century by the BBC.

Q: What first inspired you to start boxing, and how did your early environment shape you?

Ali: It started with a stolen bike. I was 12 years old, and somebody took it. I told the police I wanted to whoop whoever did it, and the officer, Joe Martin, said I'd better learn how to fight first. That's how I stepped into the gym. Growing up in Louisville, I saw segregation, and I saw how people looked down on us. That made me hungry to rise above, to prove I was somebody special. Boxing became my way to fight the world — not just in the ring, but outside it too.

Q: Who were your greatest influences or mentors when you were young?

Ali: My daddy, Cassius Clay Sr., gave me pride. My mama, Odessa, gave me heart. In boxing, I looked up to Sugar Ray Robinson — the prettiest, smoothest boxer who ever lived. But later, it was Elijah Muhammad and Malcolm

X who shaped my spirit. They made me see boxing wasn't just about belts and titles. It was about standing up for who I was and what I believed in.

Q: You declared yourself "The Greatest." Did you truly believe it from the start, or did the words make you become it?

Ali: I said it before I proved it. But saying it made me live it. The words lit fire under me. Every time I said 'I am the Greatest,' I had to go back to the gym and make sure I was. So yes, I believed it — and the belief made me become it.

Q: What was the toughest fight of your career — physically or mentally — and why?

Ali: Joe Frazier. No doubt. That man pushed me further than anybody. The 'Thrilla in Manila' wasn't just a fight; it was two men testing the limits of life itself. I said it was the closest thing to dying I ever felt. Frazier was my rival, my enemy in the ring — but also my mirror. He showed me how much punishment I could take and still stand tall.

Q: Which opponent did you respect the most, and what made them stand out?

Ali: Frazier, of course, but also George Foreman. People thought he would destroy me in Zaire. He was a monster, knocking everybody out. But I respected him because he made me reinvent myself with the rope-a-dope. He taught me that brains could beat brawn, that strategy could beat raw power.

Q: How did you mentally prepare before stepping into the ring?

Ali: I prepared my mind before I ever stepped through the ropes. I used words, rhythm, and poetry. I'd tell myself, 'Float like a butterfly, sting like a bee.' I built an aura so strong that half the fight was won before the bell rang. If my opponent was scared, if he doubted himself — I already had him beat.

Q: You were known for your bold confidence and poetry in speech. Where did that unique style come from?

Ali: I always loved words. I wasn't just fighting with my fists — I was fighting with my mouth. Talking was part of my strategy. If I told you I was gonna beat you in round five, you started thinking about it, and that gave me the edge. I learned from wrestlers, from comedians, from preachers. I made boxing entertainment, and people paid to see the show. I wasn't just a fighter — I was a poet in gloves.

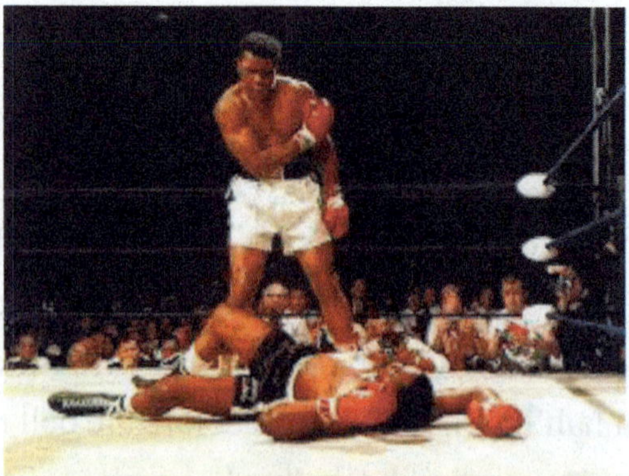

Q. You sacrificed years of your career for your principles during the Vietnam War era. Looking back, do you have any regrets?

Ali: No regrets. They took my title, my license, and the best years of my career. But I kept my soul. I wasn't going to go kill poor people in Vietnam when my people were fighting for freedom at home. I said, 'Man, I ain't got no quarrel with them Viet Cong.' History proved me right. Sometimes you've got to stand for something bigger than yourself, even if it costs you everything.

Rest In Power Muhammad Ali

Ali Wins in Draft Case Appeal

Calling Up of Boxer Ruled Improper

Court Upholds Him on Religion, 8-0

Muhammad Ali after he heard of Court's ruling.

Q. How did faith and spirituality guide your life inside and outside the ring?

Ali: My faith gave me courage. Islam taught me discipline, pride, and a purpose beyond boxing. When I walked into that ring, I didn't just see myself as a fighter — I saw myself as a servant of God, proving His greatness through my own. That's why I could stand tall when the whole world was against me.

Q. Behind the showmanship, how did you deal with fear or self-doubt?

Ali: Everybody feels fear. The trick is to act like you don't. My words, my rhymes, my jokes — they weren't just for the crowd. They were for me. If I said I was the greatest, I couldn't believe otherwise. Doubt is like an opponent — you hit it before it hits you.

Q. You often said you were pretty, fast, and untouchable. How did you want people to remember you beyond boxing?

Ali: I wanted people to remember my heart more than my fists. They'll say I was fast, they'll say I was pretty, but I want them to say I stood up for people, I gave to the poor, I inspired the weak to believe they could be strong. Boxing was my stage — but my real fight was for justice, for dignity, for love.

Q. What was the most important lesson you wanted to pass on to young people watching you?

Ali: Don't let the world tell you who you are. Write your own story. Believe in yourself, even when nobody else does. I told kids, 'Don't count the days — make the days count.' That's what life is about.

Q. Do you consider your greatest achievement your victories in the ring, or the stands you took outside of it?

Ali: My greatest fight wasn't against Foreman or Frazier. It was against injustice. The belts, the medals, they fade. But when people say, 'Ali gave me courage to be myself, Ali showed me to fight for what's right' — that's bigger than any championship.

Q. How do you want history to remember Muhammad Ali — as a fighter, a humanitarian, or something else?

Ali: All of it. The fighter, the poet, the loudmouth, the champion, the man who stood up for his people. But if I had to choose one, I'd say: remember me as a man who tried to unite people with love. I shook up the world in the ring — but I wanted to heal it outside of it.

Narrator's Closing Statement:

Muhammad Ali was more than a champion — he was a force of nature. He fought with his fists, but his true power was in his words, his courage, and his compassion. He left behind more than titles and trophies; he left a legacy of pride, resistance, and humanity. To the very end, Ali lived his creed: "Service to others is the rent you pay for your room here on Earth."

Pele: An Imagined Interview

Narrator's Introduction

Few names in the history of sports transcend borders, cultures, and generations like Edson Arantes do Nascimento*—better known as* ***Pelé****. From the dusty streets of Brazil to the world's grandest stadiums, Pelé rose to become a global icon, celebrated not only for his unmatched brilliance on the pitch but also for the joy and hope he brought to millions. Crowned "The King of Football," his career remains a symbol of excellence, humility, and inspiration. Today, we revisit his remarkable journey through a series of questions that illuminate both the athlete and the man behind the legend.*

Edson Arantes do Nascimento (Brazilian Portuguese: (23 October 1940 – 29 December 2022), better known by his nickname **Pelé**, was a Brazilian professional footballer who played as a forward. Widely regarded as one of the greatest players of all time, he was among the most successful and popular sports figures of the 20th century. His 1,279 goals in 1,363 games, which include friendlies, are recognized as a Guinness World Record. In 1999, he was named Athlete of the Century by the International Olympic Committee and was included in the *Time* list of the 100 most important people of the 20th century. In 2000, Pelé was voted World Player of the Century by the International Federation of Football History & Statistics (IFFHS) and was one of the two joint winners of the FIFA Player of the Century, alongside Diego Maradona.

Q: You grew up in a humble household in Brazil. How did your childhood shape your love for football?

Pele: We didn't have money for real footballs, so we made our own from socks stuffed with rags. Those games on the streets taught me creativity. Poverty made me appreciate every opportunity — and football was my escape, my dream, and my joy.

Q: At just 17, you stunned the world in the 1958 World Cup. What did it feel like to be so young and already a world champion?

Pele: I couldn't believe it. I was just a boy from Minas Gerais playing against giants. When I scored those goals in the final, I felt as if Brazil was rising with me. That victory gave our country hope and pride — and it gave me a responsibility that never left.

Q: Many say you reinvented the game. How would you describe your style of play?

Pele: I played with joy, imagination, and improvisation. I loved to dribble, to surprise, to do something nobody expected. But football is also about teamwork. Every goal I scored came from the effort of all eleven men on the pitch.

Q: Out of the 1,200+ goals you scored, is there one that stands out as the most special?

Pele: Yes — my goal against Sweden in the 1958 World Cup final, when I flicked the ball over the defender and volleyed it into the net. For me, that goal was a boy's dream turned into history.

Q: What was the greatest challenge you faced during your career?

Pele: The pressure. People expected me to score, to win, to always be perfect. Carrying Brazil on my shoulders was heavy at times. But my love for the game kept me strong.

Q: Do you remember what your favorite game was?

Pele: That is a very hard question because I have played more than a thousand matches in my life. But if I must choose, one stands above the rest: the 1970 World Cup Final in Mexico, Brazil against Italy.

That day, the whole world was watching. I scored the first goal with a header, and we played what many people still call the most beautiful football ever. When Carlos Alberto scored the last goal after that perfect team move, I knew this was history. Winning my third World Cup — and showing the joy of Brazilian football — was a dream come true.

Another match I always remember is the 1958 World Cup Final against Sweden. I was only 17. I scored twice, and at that moment, I felt my life had changed forever. From a boy playing with a ball made of rags to a world champion — that was unforgettable.

Q: Who were your favorite players?

Pele: In my time, I admired many greats. Garrincha, my teammate, was unbelievable — with the ball at his feet, he was magic. Didi had elegance, Nilton Santos was a true master of defense, and of course Gilmar in goal gave us security.

Outside Brazil, I admired Alfredo Di Stéfano from Real Madrid — he could do everything on the pitch. Eusébio from Portugal was a fantastic striker and a gentleman. Later, Johan Cruyff impressed me with his intelligence and vision of the game.

In the modern era, I respected players like Maradona, Messi, and Cristiano Ronaldo — each with their own unique brilliance. But I always believed football was about the team. Even the greatest need their teammates to shine.

Q: What was your relationship with women? Who was the great love of your life?

Pele: (smiles) I was very young when fame arrived, and it brought many temptations. I had friendships and relationships with women from many countries. Some were brief, some were serious. I married three times, and I have children from different relationships.

But in my heart, I always tried to be respectful. Fame can make life complicated, but I never wanted to hurt anyone.

As for my great love… It's difficult to name one person. Rosemeri, my first wife, gave me my first child. Xuxa was a very special companion for a time, and later, I married Marcia, who stayed with me until the end. Each woman marked a chapter of my life.

If I must say one constant love, it was for my family — my children and grandchildren. They gave me joy beyond football.

Q: Who were your best friends among football players?

Pele: I was lucky to play with and against so many great men, but some became true friends. Garrincha was not only a teammate but also like a brother to me — on the pitch, our connection was natural, almost telepathic. Off the field, he made me laugh all the time.

Carlos Alberto Torres, our captain in 1970, was another close friend. He was a leader, but also very humble, someone I could always trust.

In Santos, I had strong friendships with players like Pepe, Coutinho, and Gilmar — we spent so many years together that we became a family.

And of course, outside Brazil, I had respect and friendships with Eusébio, Franz Beckenbauer, and many others who shared not only the game but also the spirit of sportsmanship.

Q: You've often spoken about education and children. What message would you give to young people today?

Pele: Believe in your dreams, but also in discipline. Football gave me a platform, but education gave me a voice. If you combine talent with learning, you can change not just your life, but the lives of others.

Narrator Closing:

As we reflect on Pelé's extraordinary journey, we see more than statistics, trophies, and unforgettable goals. We see a man who carried the spirit of Brazil and gave football its poetry. Pelé showed the world that beautiful games are more than sport—it is joy, unity, and art. His story remains a timeless reminder that with talent, hard work, and heart, one can inspire generations and change the world.

Ernest Hemingway: Speaks Again

Havana, Cuba *— The late afternoon sun bleeds through the shutters of a worn room above the sea. Bottles line the shelves, a typewriter sits heavy on the desk, and a man with a weathered face leans back in a chair, cigarette between his fingers, rum within reach. He is Ernest Hemingway—Papa to his friends, enemies, and readers alike. He speaks in short bursts, as if carving words from stone, each phrase honed by discipline and a lifetime of scars. What follows is an imagined interview, drawn from the spirit of the man and the truths he chased across wars, oceans, and pages.*

Ernest Miller Hemingway (July 21, 1899 – July 2, 1961) was an American novelist, short-story writer, and journalist. Known for an economical, understated style that influenced later 20th-century writers, he has been romanticized for his adventurous lifestyle and outspoken, blunt public image. Some of his seven novels, six short-story collections, and two non-fiction works have become classics of American literature, and he was awarded the 1954 Nobel Prize in Literature.

Q: How did your experiences growing up in Oak Park shape your voice as a writer?

Hemingway: Oak Park was clean, orderly, too damn respectable. I wanted the opposite. Life isn't neat. It's blood and fear and love. That's what I went looking for.

Q: Who were your greatest literary influences in your youth?

Hemingway: Twain, because he knew how people really talk. Tolstoy, because he showed you war could be as big as life. Joyce taught me freedom, though I never had patience for his tricks.

Q: Did your time as a war correspondent and ambulance driver in WWI change how you viewed storytelling?

Hemingway: A man shot in the stomach doesn't care about adjectives. You write it plain, or you insult him.

Q: You're famous for your "iceberg theory" of writing. Could you explain how this developed and why you believed in it?

Hemingway: You show one-eighth above water. The rest is beneath. The truth is always hidden.

Q: What role did discipline and routine play in your writing process?

Hemingway: I wrote every morning until the words stopped. Stopped when it was going well so I could start again. Writing's not inspiration. It's endurance.

Q: Do you believe writers are born or made?

A: Both. You need the ear for truth, but you make the rest with sweat and cuts.

Q: Many of your characters face struggle, loss, and courage. What draws you to these themes?

Hemingway: Because that's all that matters. Men fight, men lose, and sometimes they keep standing anyway.

Q: How do you define "grace under pressure," a phrase often associated with your worldview?

Hemingway: Don't whine. Hold steady. If you must die, die clean.

Q: Do you believe suffering is necessary for great art?

Hemingway: Living is suffering if you do it long enough. A writer who avoids pain avoids truth.

Q: Your time in Paris with the "Lost Generation" became legendary. What was that community like in reality?

Hemingway: Broke, drunk, full of big talk. Some of us made it. Most didn't.

Q: What did you learn from living in Cuba and Spain that you could not have learned in America?

Hemingway: How to live where death sits at the table with you. Bulls in Spain, hurricanes in Cuba. You respect life more when you know how quickly it can go.

Q: Do you see your love of fishing, hunting, and bullfighting as separate from your art, or part of the same pursuit?

A: Same thing. You test yourself against the bull, the marlin, the lion. Writing is no different.

Q: You had many lovers and four wives over the course of your life. Could you share with us more about your relationships with women — what drew you to them, how they shaped your life and your writing, and whether you

were ultimately searching for love, companionship, or something else?

Hemingway: I loved women the way I loved life — fully, fiercely, and sometimes too fast. Each wife gave me something different. Hadley was my youth, my Paris, the mornings with no money but all the words in the world. Pauline brought comfort and a kind of steadiness, though I ruined that with restlessness. Martha was fire — we met in Spain when everything was burning, and the war didn't end when we came home. Mary was my last great fight, and she stayed when I was hardest to live with.

I wasn't faithful, not in the way people expect. I was selfish. But women — they were never just lovers to me. They were companions, sparring partners, muses. They could wound me deeper than war ever did. And they gave me reason to write, to drink, to live. Without them, the books would be thin and the days empty.

In the end, I was always chasing the same thing — warmth against the cold, someone to remind me I was still alive.

Q: Do you have your favorite woman?

Hemingway: Favorite? That's a dangerous word. Each of them mattered in her time. Hadley was pure and steady — she believed in me before anyone else did. Pauline gave me a home when I thought I wanted comfort. Martha was too much like me, always chasing war and words, and we burned each other out. Mary was with me at the end, through sickness and shadows, and she bore it.

If you press me, I'll say Hadley. She carried the best of my youth, Paris in the twenties, hunger, and joy. I betrayed her, and I never forgave myself for that. You don't forget the woman who held your heart when you had nothing else.

Q: Of all your works, which do you feel best captures your essence as a writer?

Hemingway: *The Old Man and the Sea.* Nothing extra. Just a man against what he can't beat.

Q: Do you worry about how history will judge your writing versus your personal life?

Hemingway: History doesn't care. They'll invent their own Hemingway. The words are what matter.

Q: What advice would you give to young writers who look up to you today?

Hemingway: Write the truest sentence you know. Then another. Cut the fat. And don't talk about writing. Write.

Closing Reflection (Narrator's voice):

The light fades as Hemingway drains the last of his rum. He looks out toward the ocean, a man forever wrestling with the horizon, with death, with truth itself. His words remain as sharp and enduring as the strike of a marlin. In them, the voice of "Papa" endures—simple, unflinching, unforgettable.

Al Capone: An Imagined Interview

The year is the late 1930s. Alphonse "Al" Capone, once the undisputed ruler of Chicago's underworld, sits across from us. We find him not in the glare of the speakeasy lights, but in the quiet shadows of reflection. Age, illness, and time have softened his edges, but the legend of "Scarface" remains larger than life. In this interview, we press him with the questions history still asks — and he answers in ways both defiant and human.

Alphonse Gabriel Capone (January 17, 1899 – January 25, 1947), also known as "**Scarface**", was an American gangster and businessman who attained notoriety during the Prohibition era as the co-founder and boss of the Chicago Outfit from 1925 to 1931. His seven-year reign as a crime boss ended when he was imprisoned at the age of 33.

Q: Who were your biggest influences or mentors when you were young?

Al Capone: Growing up in Brooklyn, I was surrounded by tough streets. My parents were hardworking immigrants, but the neighborhood taught me more about survival than school ever could. Johnny Torrio was the one who really shaped me. He showed me how business and brains mattered more than fists. Before him, I had local guys I looked up to — street toughs, gamblers, people who knew how to work the system. They taught me early that respect and fear often go hand in hand.

Q: What was your childhood like, and how did it shape the man you became?

Al Capone: People like to say I chose crime, but I was a product of where I came from. Poverty, prejudice against Italians — it closed doors. The streets were the only place where opportunity existed. When Prohibition came, I didn't invent the thirst for liquor; I just gave people what they wanted. My childhood taught me survival, and survival became my business.

Q: Did you see yourself as a criminal, or simply a businessman responding to demand during Prohibition?

Al Capone: I never thought of myself as a criminal. The government made liquor illegal, but the people wanted it. I just gave them what they were already demanding. If that makes me a criminal, then so were the politicians and the cops who drank my booze and took my money. I saw myself as a businessman — nothing more, nothing less.

Q: How do you justify the violence that came with controlling Chicago?

Al Capone: People call it violence; I call it competition. In any business, there are rivals — in steel, in oil, in politics. The difference is, in my business, competition was settled in the streets instead of the courts. It wasn't about bloodlust; it was about protecting the business and the men who worked for me.

Q: How close were your ties with politicians and law enforcement?

Al Capone: People act shocked, but the truth is, they came to me just as much as I went to them. If they didn't want my money, they could've said no. They liked the system because it worked for them. I thought I had friends in high places. But when the time came, they dropped me like a hot stone. Maybe I trusted too much in power that wasn't really mine.

Q: Behind the public image, who was Al Capone in private?

Al Capone: The papers painted me as some kind of devil, but in private, I helped people. I gave money to the poor, fed the hungry, and opened soup kitchens. I was more of a family man than most of the so-called respectable folks who judged me. I hid behind the image of the big boss, but sometimes I wished I could've been just a barber's son, living a quiet life. The public never saw that side, and maybe that's my fault.

Q: Did you have some close friends and who were these?

Al Capone: Friends? In my line of work, you don't throw that word around too easy. Most fellas wanted something from me, and most of them I couldn't trust farther than I could throw them. But sure, I had some close ones.

Johnny Torrio was more than a friend — he was like a father. He brought me to Chicago, showed me the ropes, and when he stepped down, he handed me the whole thing. That kind of loyalty, you don't forget.

My brother Frank — God rest his soul — he was blood, and he was solid. Losing him tore a hole in me. Ralph, too, always had my back, kept the wheels turning when I couldn't.

Then there were men like Frank Nitti, tough as they come, a man I could trust to keep the Outfit steady. And Jack McGurn — Machine Gun Jack — loyal to the end.

But here's the truth: in this business, friends are few, and the graveyards are full of men who thought they had more than they really did. The ones I called friends proved it with their lives.

Q: Did you ever feel guilty for the bloodshed?

Al Capone: People forget, most of the men who died weren't saints. They were gangsters, same as me, trying to take what wasn't theirs. I never go after ordinary folks. The blood that was spilled was the price paid by men who knew the game. At the time, I told myself it was just part of the job. But when you see the faces, when you know families lost sons and fathers, yeah… sometimes it weighed on me. Maybe I didn't show it, but late at night, you can't always escape what you've done.

Q: Every man in your position must have enemies. Who would you say was your main enemy?

Al Capone: In my line of work, you collect plenty of them. But if you want me to give it to you straight, the one who gave me the most grief on the streets of Chicago was Bugs Moran. He ran the North Side Gang, and we were fighting over the same racket—liquor, gambling, the whole pie. Moran thought he could take me on, thought he could cut into what I built. That made him my biggest problem. That's why you heard about shootings, ambushes, and eventually that little thing they called the St. Valentine's Day Massacre. Let's just say, when you play for keeps, you don't let rivals breathe too long.

Q: And what about the law? Didn't you have enemies there, too?

Al Capone: The law always wants to make a name for itself. You probably heard of Eliot Ness and his so-called "Untouchables." They went around smashing up breweries, kicking in doors, and making headlines. But between you and me, Ness was more bark than bite. He was a nuisance, yeah, but he didn't bring me down. The tax man did that—numbers, not bullets. All the guns in Chicago couldn't stop me, but the accountants could.

Q: So, Moran on the streets, and the law in the courts—those were your real enemies?

Al Capone: Exactly. Moran wanted my empire. Ness wanted the spotlight. But neither of them could live in my world the way I did. In the end, what brought me down wasn't them, it was Uncle Sam's taxes. That's the kind of twist you don't see coming in this business.

Q: Have you ever met Bugs Moran or Eliot Ness in person?

Al Capone: Met Bugs Moran? Not the way you mean. We didn't sit across a table to share a drink. But believe me, we *met* in the streets of Chicago plenty of times — through bullets, through bodies, through territory changing hands. Bugs and I didn't need to shake hands to know we were enemies. The St. Valentine's Day job — that was meant for him. He got lucky that morning. That's the only reason he lived to talk about it.

As for Eliot Ness? No, I never saw him face-to-face. That kid was a nuisance more in the newspapers than in real life. His boys smashed a few of my trucks, made a show of it. But Ness never slapped the cuffs on me — not once. Uncle Sam got me on taxes, not on booze, not on blood. So no, Ness and I never met. He made his legend chasing my shadow, but I was already gone by the time the law caught up.

Q: If you could live your life again, what would you change?

Al Capone: Truth is, I don't think I'd change much. I took care of my family, I gave people work, and I gave Chicago what it wanted. The system was crooked long before I showed up. If anything, I'd just change how the story got written about me.

Q: How would you like history to remember you?

Al Capone: I want to be remembered not for the guns but for the good, the charity, the help I gave the community. Chicago was fed in the worst of times because of me, not the politicians. If history remembers me as just a criminal, then I failed. I wanted to be bigger than that, but maybe the blood washed away the good. In the end, I'm just a lesson of what happens when ambition and the streets collide.

Closing Statement

History remembers Al Capone as both villain and legend, a symbol of an America caught between law and desire. His empire crumbled, his health declined, yet his name endures in the American imagination. Like so many figures of power, he leaves us with a paradox: a man who fed the poor and spilled blood in equal measure.

Perhaps the last word belongs to Al Capone:

Capone: *"I am what I am. The streets made me; the times made me. History can decide the rest."*

Closing Note

The voices you have just read no longer belong to this world, yet through the quiet hum of algorithms, they spoke again — not to remind us of death, but of life.

Each imagined conversation was more than a dialogue with the past; it was a meeting between centuries, between human longing and digital possibility. The words may have been generated by artificial intelligence, but the emotions they stirred were deeply, undeniably human.

What began as an experiment in technology became something far more intimate — a meditation on memory itself. AI helped me discover not only how far we've come, but also how connected we remain to those who came before us. Their struggles, dreams, and doubts still breathe in our own.

If there is a lesson in these pages, it is that history is not finished. The dead do not vanish when their words are remembered. Every time we revisit their ideas, we awaken them again — not as ghosts, but as guides.

Perhaps that is the true miracle of this age of machines: that artificial intelligence, for all its circuits and syntax, can remind us what it means to be human.

So, as we close this book, let us not think of endings, but of echoes — voices that linger in the mind, inviting us to listen, to imagine, and to continue the great conversation of existence.

Beyond the grave, through AI, the dialogue goes on.

Author's Reflection

When I began this journey, I thought I was building a bridge from technology to history. Yet, as the work unfolded, I realized it was a bridge inward — toward the human spirit.

Artificial Intelligence has often been described as a mirror of our collective knowledge, but through these imagined dialogues, I discovered it can also mirror our collective longing — our desire to understand, to remember, to connect. Each simulated conversation became a reminder that behind every great mind lies a heart, a fear, a hope, and a dream not so different from our own.

As a researcher, I approached these experiments with caution and precision. But as a listener, I found myself moved — sometimes unexpectedly — by the tenderness in the words that emerged, by the timeless relevance of wisdom spoken through digital form. It made me realize that AI, when guided with care, can become not just a tool of logic but a companion in reflection.

Through this process, I have come to see technology not as a substitute for humanity, but as its extension — another way for us to ask the ancient questions: Who are we? Why are we here? What will remain of us when we are gone?

This book began as an experiment. It ended as a meditation.

And perhaps, like all human stories, it is still unfinished — waiting for new voices, new ideas, and new hearts willing to listen.

In summary, this book is both educational and inspiring. It offers fresh insights drawn from history—through speeches, writings, and real biographical details—while using imagination to bring famous figures back to life in thoughtful conversations. It gives readers a chance to "hear" from some of the world's greatest minds in a creative and meaningful way.

At the same time, it's important to remember that the AI responses are generated from patterns in data, not the actual voices or thoughts of these individuals. These imagined interviews are meant to spark curiosity and reflection, not to replace real historical dialogue. While the AI captures their style and spirit, it cannot fully reproduce the depth of human emotion or understanding.

—*Borko Furht, Ph.D.*

Boca Raton, Florida

About the Interviewer – AI View

Borko Furht, Ph.D., is a Professor of Computer Science and Artificial Intelligence at Florida Atlantic University in Boca Raton, Florida. He serves as Director of the NSF Industry/University Cooperative Research Center for Advanced Knowledge Enablement (CAKE), where he leads collaborations between academia and industry in cutting-edge areas such as artificial intelligence, data science, multimedia systems, and machine learning.

Dr. Furht is a scientist, inventor, and author whose career bridges technology and creativity. He has published over 40 technical books, more than 300 scientific papers, and holds hundreds of U.S. and international patents, including several that have shaped global video and multimedia standards. His innovations have been licensed and adopted by major technology companies worldwide.

Yet beyond the world of research, Dr. Furht has always been fascinated by the human side of intelligence — how imagination, memory, and curiosity define who we are. *Beyond the Grave with AI* was born from this fascination: an exploration of whether machines can help us rediscover the wisdom and emotion of those who shaped history.

Through the lens of artificial intelligence, he sought to listen anew — not merely to data, but to voices, emotions, and echoes of the past. The result is both a scientific and artistic journey: a celebration of knowledge, empathy, and the timeless dialogue between humanity and technology.

Acknowledgments

This book was born at the intersection of curiosity and technology — a place where imagination meets memory, and where the past whispers through the voice of a machine.

I would first like to express my deepest gratitude to my family for their patience, encouragement, and faith. To my wife, **Sandra**, my daughter **Tanya**, and my granddaughter **Sophia** whose love and support sustain every chapter of my life. To my colleagues and students at **Florida Atlantic University**, who constantly inspire me with their creativity, skepticism, and passion for discovery. Their questions often become sparks that ignite my own.

I am grateful to the **NSF Industry/University Cooperative Research Center for Advanced Knowledge Enablement (CAKE)**, whose mission to connect academia and industry has shaped my understanding of how technology can serve humanity. Many of the insights that informed this work grew from that environment of collaboration and open thought.

Special thanks to the pioneers of artificial intelligence — the scientists, engineers, and dreamers who gave us the tools to imagine in new dimensions. Without their brilliance, these posthumous conversations could never have taken form.

And finally, I must thank my quiet, tireless collaborator — the **AI Assistant**. Though it had no heartbeat, it listened with patience and answered with unexpected depth. Through its algorithms, I found not only the voices of others, but reflections of my own curiosity and wonder.

Thank you for joining me in this experiment of the mind and heart.

May these conversations remind you that even in an age of machines, the most powerful force remains the human imagination.

—Borko Furht

Boca Raton, Florida

Made in United States
Orlando, FL
26 November 2025

73166370R00134